The Making of a Scientist

THE MAKING
OF A SCIENTIST

BY ANNE ROE

GREENWOOD PRESS, PUBLISHERS
WESTPORT, CONNECTICUT

Library of Congress Cataloging in Publication Data

Roe, Anne, 1904–
    The making of a scientist.

    Reprint of the ed. published by Dodd, Mead,
New York.
    1. Scientists. I. Title.
Q141.R52 1973   502'.3   73-15059
ISBN 0-8371-7151-2

The poem "Homily for Art Students," appearing on page 11, copyright, 1946, by The New Yorker Magazine.

Designed by Stefan Salter

Originally published in 1953 by Dodd, Mead &
Company, New York.

Reprinted with the permission of Anne Roe.

Reprinted in 1973 by Greenwood Press
A division of Congressional Information Service, Inc.
88 Post Road West, Westport, Connecticut 06881

Library of Congress catalog card number 73-15059
ISBN 0-8371-7151-2

Printed in the United States of America

10 9 8 7 6 5 4 3 2

For My Favorite Scientists:

MY HUSBAND AND MY SUBJECTS

Contents

Tables

Figures

Photograph

The Making of a Scientist

I

How This Research Study Developed

THIS IS the story of four years of research on the most fascinating problem in the world,—at least it seems so to me. The problem is what kinds of people do what kinds of scientific research and why, and how, and when. In attempting to study this I necessarily became acquainted with many of the leading research scientists in the United States. I found out about their families, about how they had learned about science and how and why they had become scientists. I gave them intelligence and personality tests, and I read and tried to understand their work. I worked with biologists, with physical scientists and with social scientists, that is psychologists and anthropologists. This is how I went about it and what I found.

I am a psychologist. All psychologists are concerned with the behavior of people or of animals. (Those who study animals do so in large part for the light this may throw on the more complicated behavior of humans.) There are a number of different kinds of psychologists who do a number of different things. For example, experimental psychologists tend to study one particular process, such as vision or learning, and the circumstances which aid or hamper the process. Clinical

1

psychologists are interested in the person as a whole, in what makes him tick, and in helping him when something goes wrong. Social psychologists are interested in how people interact with each other and the behavior of people in groups. There are also industrial psychologists, school psychologists and so on.

I am a clinical psychologist although it is a long, long time since I have worked in a clinic. Clinical psychologists are so called, I suppose, because when this branch of psychology was developing they mostly worked in mental hygiene clinics or child guidance clinics, or in hospitals, and many of them still do. The thing that distinguishes them from other psychologists, however, is that their primary concern is with the total individual as a functioning person. Of course the clinical psychologist relies heavily upon data supplied by experimental and social psychologists.

Clinical psychologists usually spend most of their time with people who need help for one reason or another. Few of them have had or have made the opportunity to do intensive clinical studies of normal people. Of course they have used such groups as college students, or other "captive audiences" for the standardization of the various tests which are now so popular, but these are not clinical studies. By a clinical study is meant a study of all aspects of the person in a related whole. A thorough clinical study would include life history, the level of intellectual and other capacities, the adequacy of their functioning, and the structure of the personality, that is, what motivates the person, what needs he has and how he satisfies them. In psychological jargon, the person who takes part in such a study, or in an experimental study is called a "subject" (perhaps because he is subjected to so many annoyances)!

Many clinical psychologists do some research but few can spend all their time at it. For me, as for many others, research is more fun than anything else, so I consider myself particularly fortunate that almost all of my professional life has been spent in full-time clinical research of various sorts, and a very large amount of that with normal people. (My definition of a normal person is one who follows the usual pattern of his social group without undue difficulty and without needing special help. This does not mean that he has no problems, perhaps even serious ones.) I have studied the intelligence of average adults and of adults with various mental illnesses. I have studied aphasics (people with a particular variety of speech disorder), problem children, new-born infants, foster children, alcoholics and painters, but I have never done anything half so exciting as studying research scientists.

Not only exciting, but also very worrying at times. People are sensitive, even scientists, or perhaps especially scientists, and you never know what you will run up against. They did not need my help, I was asking for theirs, and it was certainly not my function to upset them, but any thorough personal study is likely to have its upsetting moments if it is any good. This makes the psychological research worker's job a very ticklish one. I always had at least a few moments of panic before tackling each scientist for the first time. If some of my subjects were apprehensive before seeing me, I was at least as apprehensive as they. Even so I did not fully realize what an incredible nerve the project had taken until I had completed it.

I had gotten myself into it—no one had asked me to do it. What I had done was to get hold of sixty-four of the country's leading scientists, get them to give me literally hours of their time, during which they told me about themselves, and let me give them the works, psychologically speaking. That they were willing to cooperate in such a project has various explanations. Some were quite simply curious as to what psychologists were up to, and curious enough to give a good deal of time to find out. (Curiosity about lots of things is a characteristic of a first-class scientist.) It is always interesting to talk about oneself, too, and it is flattering to be picked as one of the "most eminent." But most were motivated primarily, I think, by the realization that the results could be important to their own sciences, as well as to psychology, and scientists themselves, they could be detached enough to become guinea pigs for a time.

How this particular project developed, and how I went about getting it under way are fairly typical of any sort of scientific research at the present time. I say at the present time because the position of science and of scientists changes with more general changes in the social structure as well as with technical advances. Not only would this particular project not have been possible some years ago in this form, but the means for doing it, if they had come to hand, would surely have been different.

Any scientific study develops out of the work that has preceded it (whether of the same or of other workers) and the particular predilections of the individual worker as far as he is in a position to control his own research. Research scientists usually are in a position to control their own work and this is one of the reasons, I think, why they derive such tremendous gratification from it. But there are special situations in which the scientist does not have so much freedom. Scientists in industry rarely have it, and scientists working as members of a team are also constrained, at least as far as the problem to be investigated is concerned. In the special circumstances of mobilization of scientists for war, again the problem is set for them, although some of them, at least, must have considerable freedom in working out the methods of approach. Any diminution of a scientist's accustomed freedom to control his own work and what he says about it is a major frustrating experience to him and one for which he naturally has very little tolerance. All of his professional activities are predicated, in a sense, upon the possession of this freedom. Normally, however, the scientist is his own boss as far as his research is concerned, and he is limited only by the time at his disposal (his primary job may be college teaching, for example) and the equipment available to him (although this is less often a serious limitation than one might think).

This study of scientists derives very directly from an earlier study of mine, of artists. Up to that time, while there had been many studies of vocational aptitudes, or special skills related to particular vocations, there had been no clinical studies of vocational choice or performance in terms of life history or personality structure. Nor had there been any clinical studies of highly successful people of any sort (except of a few who had needed psychological or psychiatric assistance).

The primary aim of the earlier study was quite different from that of this one. At that time I was on the staff of the Yale School of Alcohol Studies which has been engaged in studying all sorts of things about the use of alcohol, from its immediate physiological effects to its social consequences. I had just finished a study of how children of alcoholic parents turn out when compared to children of normal parents, if all of them are raised by foster parents. (They turn out very well.) It then had to be decided what I should undertake next, and since I was a psychologist working within the framework of a specialized organization, it had to be a psychological study which had something to do with drinking. There were innumerable possibilities, of course. The director, E. M. Jellinek, suggested that it would be of very great interest to investigate the problem of the relationship between alcohol consumption, and such a creative activity as painting. The literature is full of statements and speculations about such a relationship, but there had been no direct work on it. He proposed that this should be a bibliographic study, that is that I should read a good many biographies of famous painters, and compile all the data I could find on the subject.

Here is an example of how the personal predilections of the individual worker come in, because while the problem interested me very much, the proposed approach not only bored but repelled me. This is really a personal matter, because some very good work of this sort has been done, such as the study of the intelligence of men of genius of the past which Catherine Cox did. Estimates of intelligence, however, are a rather different matter from estimates of drinking habits, because of the moral and emotional attitudes toward alcohol. There would be many pitfalls. Even first-hand accounts are subject to various intentional or unintentional distortions, and second and third and fourth hand ones, with innumerable possible and probably unascertainable biases on the part of the biographer seemed almost impossible of assessment in any scientific fashion. I therefore made the counter proposal that, instead of reading books about painters I hunt up a number of the most eminent American painters and ask them about themselves. This would obviously involve a number of difficulties, as well as cost considerably more than a library study, but it was readily agreed to if, as was thought doubtful, I could get away with it.

Then it seemed to me that if I was going to get hold of such a group of people, it would be sensible to find out a lot more than just their drinking habits and what effect, if any, this had on their painting. The use of psychological tests would not only be in itself a highly desirable contribution to our knowledge of "the creative personality," but giving the tests to established artists would serve as a check on the tests themselves. The tests would also make it possible to evaluate the subject's account of himself more effectively. Furthermore, it is always easier to work with test scores, to summarize from more or less objective data than it is from descriptive material.

After consideration of recent developments in psychological testing, I decided to make use of two "personality" tests, the Rorschach Method and the Thematic Apperception Test, which I will describe later. These would be more pertinent to the general study than an intelligence test and besides I had a strong suspicion that it would be difficult if not impossible to persuade these subjects to take an intelligence test, whereas taking the other tests might well amuse them. I had not used either of these tests before so I had first to spend a considerable amount of time learning how to use them properly; they put a much greater burden on the psychologist than almost any other test. Then it was necessary to acquaint myself with modern American art and artists, and as I had to do this from scratch, this also took some time.

The classical scientific experiment is one in which a single factor is varied, all others remaining constant, so that the effect of this single factor on the others can be studied very clearly. In psychology it is impossible ever to isolate a single factor, and even in physics this simple set-up is rarely still believed possible. Instead one must study simultaneously the interaction of a number of factors. This is not too difficult, with modern methods of statistical analysis, when all of the factors involved can be identified and measured. In psychology, again, very often neither of these things can be done with certainty. For these reasons, even experimental psychological research cannot fit the old classical model, and attempts to force it to do so are likely to vitiate the work, since most often they just result in the experimenter's ignoring relevant variables. The research I am reporting here is not experimental but clinical research. This means that instead of my setting up a situation and inducing my subjects to do particular things for me, the situation is already set up for me, and the subjects have already been behaving in whatever ways are natural to them. At this stage the procedure must be largely an observational one; later it may be possible to set up hypotheses and check them on other groups on the basis of what is learned about this one.

In clinical research, as well as in experimental, it is obviously simpler to avoid the introduction of any more variables into a problem than is absolutely necessary. We know that some intellectual functions change with age, and that there are sex differences in certain psychological variables. For this reason I limited my study of artists to men and to men mature enough to have achieved eminence but not so old that age in itself might affect the situation seriously. A further limitation, that the men should live in or near New York City was imposed by the extra cost of visiting those who lived at a distance. This did not seriously affect the group as representative of American painters since there is a marked concentration of them in that area. When I started out I knew nothing about the relation between style of painting and personality so it seemed best to include representatives of many different styles of painting. For the specific purposes of the study, of course, the major variable was amount of drinking, and hence I wanted as wide a range in this variable as possible,—from total abstainers to excessive drinkers. (I never did succeed in finding a total abstainer, although I did find some very moderate drinkers.)

Before I could even begin to work with any individual subject, then, several months of preliminary work had been necessary. Furthermore, this was in a sense on speculation. I proposed to report the work without using the names of the subjects, but even with this safeguard, I had no assurance at all that it would be possible to obtain the subjects I wanted. Indeed I was assured more than once that no eminent artist would be willing to submit to such an extremely personal investigation. I rather thought this might turn out to be the case, but it seemed to me and to the organization with which I was working that it was worth trying. It is characteristic of scientific work that one has to gamble frequently in this way. It is often necessary to put in months of preliminary work, and it quite often turns out that the necessary techniques are not yet available, or some unforeseen gimmick develops. Then the work never gets beyond the preliminary stages and the time and expense have to be written off. I do not believe, however, that research plans that do not materialize are ever a complete loss,—the worker has always learned something which he uses later and sometimes technical advances make it possible to do the work eventually. More than once, fur-

9

thermore, a study has been completed with only negative results. It is even more true that negative results are never a complete loss. It can be just as important to know that something is not so as to know that it is. So I went ahead with this study as though I had been sure at the outset that it could be done.

The details of the final selection of this group do not matter here, but the doubters were wrong. Of the 23 artists I asked, 20 did serve as subjects and the results have long since been published in various technical journals. For those of you who are curious about what was learned about any relation between drinking and painting, I will summarize the results briefly. Only one of the painters customarily drank to help him get started on a painting. All of the others reported that, although they might paint when they were drinking, and it might seem at the time to be wonderful, the work almost always had to be discarded in the end. On the other hand, most of them felt that moderate amounts of alcohol at the end of a hard day were a decided help in relaxing and in the long run served materially to keep down the nervous tension that can build up so easily and that can hinder creative efforts. Creative work involves a very delicate balance between tension and relaxation.

There were other results from this study, of which perhaps the most interesting were the relationships which could be found between what and how the man painted, and what sort of person he was and what sort of problems he had. These relationships are very direct, once you learn to find the cues, and quite easy to demonstrate. To anticipate the study of scientists, I may say that although such relationships obtain among them also, they are usually much more obscure, and the technique of reporting scientific results serves rather to hide the man than to display him. There are greater personality differences among artists than among scientists, but successful artists and successful scientists have one thing in common,—their intense devotion to their work.

There is an amusing and accurate, if incomplete, summary of the results of the study,—a verse by Arthur Kramer that was published in *The New Yorker* a short time after a newspaper account of the study had appeared:

HOMILY FOR ART STUDENTS

Study of Painters Shows Drink Does not Help Creative Work.
—*The Times.*

> Turners, Dürers, or Bellinis
> Do not spring from dry Martinis
> Goya's genius, Rubens' powers
> Did not stem from whiskey sours.
> Fumy brandies, potent ciders
> Make no Holbeins, make no Ryders.
> Alcohol's ingurgitation
> Is, in short, no substitution
> For creative inspiration
> Or artistic execution.
>   Guzzle *vino*
>   Till you're blotto—
> Splotches will remain but splotches.
>   Perugino,
>   Ingres, Giotto
> Were not born of double Scotches.
> Nor, alas, will full sobriety
> Whisk you into their society.
>   —*Arthur Kramer.*

It was tentatively planned to follow the study of artists by similar studies of writers and composers for the School of Alcohol Studies, but at this point my husband returned from war and I returned to New York City where we lived. Of course I was very pleased that it had been possible to do the study but I was not intensely interested in the alcohol angle. I had been astounded at the extraordinary amount of information that the combination of interviews with these two tests had yielded with regard to the psychological meaning of painting as a vocation and of the significance of the individual's style of painting as well as its content. I wondered if it would not be possible to study scientists in the same way.

Particularly since their contribution during the war, the role of scientists in modern society is a rapidly changing one. The absent-minded professor, ivory-tower dweller stereotype no longer fits them, if it ever did, and this is generally realized. It is enormously important for us as a society to understand what science can and cannot be expected to do and this is even more true of its basic research than of its obvious technological applications. But science is the work of scientists, and what kind of person the scientist is and why and how he becomes a scientist had never been seriously studied. I wanted to find out. Here, too, it is doubtless true that personal predilection played a considerable part. Science and scientists have always fascinated me. I married a scientist. My personal associations are chiefly with scientists and since my husband's field is very different from mine, I meet many different kinds of scientists.

One can tell a lot about a painter from his painting, if one sees enough of it done over a long period of time. Can one deduce anything about a scientist from his work? Both groups

do creative work, in a broad sense, but are the obvious differences in the kind of work they do related to differences in
personality or early experiences, or to something else? What
does science mean to the scientist? Are scientists different
from other people? Is one kind of scientist characteristically
different from another? All these and many other questions
kept plaguing me, the more so as I felt that it was extremely
probable that the techniques I had used in studying artists
might go a long way towards getting a start on the answers to
these questions.

I actually had a little evidence on some of these points. In
another connection I had used the Rorschach with a group of
vertebrate paleontologists. (Paleontologists study the evolution and history of animals and plants through their fossil remains. Vertebrate paleontologists study the group of animals
that have backbones. The Society of Vertebrate Paleontology
has an annual meeting and one year all present were good
enough to take the Rorschach as a group.) There seemed to
be quite a characteristic pattern in many of this group of
scientists, which made a good deal of sense in terms of a relation between vocation and personality structure. This confirmed my guess that this method could give useful and pertinent information about the relationship between personality
and vocation.

After much mulling over these questions and such evidence
as was available on them I drew up a description of a proposed study. Then the problem was somehow to find ways
and means of doing the work. It would be a fairly expensive
thing to do. Scientists are scattered all over the country although there are some regional concentrations, and this meant
a good deal of travelling. I would need technical assistance

and secretarial assistance, and all the general overhead expenses that are usually part of an institutional set-up. I had no institutional connection at that time, which was a difficulty because the individual scientist is rarely given a large research grant. Such a grant is almost always administered through a university or hospital or similar institution which assumes the responsibility for handling the funds, even though full control of the project is in the hands of a single member of the staff.

I tried the two most obvious private foundations. Neither was interested. Neither gave funds to individual investigators. Furthermore neither had any tradition for the granting of funds for clinical psychological research and they saw no reason for starting with this study. It is clear that such a project as this is a much more ticklish matter than say, computing IQs for students in different fields of work.

About then the American Psychological Association was holding its annual meeting. I reported the study I had made of vertebrate paleontologists. In the course of the ensuing discussion I remarked that I had been looking in vain for funds to pursue further the general problem of the relation between vocation and personality. Several of my colleagues suggested that I apply to the United States Public Health Service, which had just set up an organization to give research assistance of various sorts. Obviously it would do no harm to try, so I obtained and filled out the forms and sent them in,— and took a job with the Veterans Administration.

Receipt of the forms was promptly acknowledged, and there the matter rested for some months, during which I became immersed in drawing up an ambitious research project for the Veterans Administration. One Sunday morning of hallowed memory, when I was indulging in breakfast in bed and reading the *New York Times*, I glanced idly through a news dispatch. The headline said something about grants for cancer research from the United States Public Health Service, but other types of grants were also listed and there I read that one had been made to me! I nearly fell out of bed. There followed two weeks of complete silence on the part of the United States Public Health Service, and finally I could stand it no longer and wrote to inquire. The director of research somewhat huffily confirmed the news story which had gotten out ahead of time.

I can admit now that I had been pretty blithe about the project up to then. I had had no real expectation of ever getting the funds to do it, but it seemed like such a good idea that I felt I had to try at least. Now I was actually confronted with having to do it and I was appalled. It would certainly be the most difficult job I had ever tackled and I was not absolutely sure that it could be done at all. It meant, too, being away from home for some months on end. True, the children were grown and needed me only intermittently but I do not like being away from my husband and like most wives I have a firm conviction that he is not as comfortable when I am away.

The career vs. family problem is one that has to be solved by every professional woman and her husband in their own fashion, and we had had to work it out, too. My husband, himself a research scientist, is well aware of the driving necessity to keep on doing research that is characteristic of all who have experienced its satisfactions, and he is willing to be discommoded,—up to a reasonable point. On the other hand, I have always felt that a successful marriage is impossible for a professional woman unless both she and her husband have a clear (and the same) conviction of what comes first, even though either one might have serious conflicts about this at times. I think it probable that a successful marriage can be worked out with the wife putting either career or family first, so long as everyone is clear about it. As for me, I have never had any doubts that I had to subordinate my own work to the welfare of the family, although in all candor I must admit that I have deeply resented it upon occasion. Now that the children are grown, and we are learning at first hand the delights of being grandparents, I am well satisfied. I do not mean that there are no further problems. There is always the problem when your husband is a man of great stature, of balancing your own work against the chances of maintaining or increasing his productivity, so as to be sure that in the long run you have not benefited at the expense of society.

Now I had the money. I had to do the job. The grant was for one year with the expectation that it would be renewed twice. I had asked for three years, but government agencies work on annual budgets, and I had planned the project so that each year's work was a unit. In the end the grant was extended for three years and nine months, by which time the obvious part of the work was done, the reports written and in press. But there is still much that can be learned from these data.

This year I have a Guggenheim Fellowship. This means that I have a modest income for a year, and no duties except to think and write about the experiences of these last few years. I have a great deal of material on the creative process itself, as exemplified in the work of these men, and this material has never been worked over. This will take time, long uninterrupted periods of brooding over it, and this I now will have time to do. I am spending this time in a remote spot in New Mexico where the nearest telephone is ten miles away, and mail is delivered five miles away three times a week, so there will not be many interruptions. There are some distractions, pleasurable ones. Mule deer come by the house frequently, and one morning I looked up from my typewriter in time to see a flock of wild turkeys busily eating their way across the field outside my window.

First, though, I am writing this book. For one thing it will enable me to review the experience as a whole. For another, it will make it possible for me to give my subjects a more coherent and comprehensible account than the technical reports give to anyone but an expert in my own field. My subjects have surely earned this, and I feel a strong sense of obligation to them. Most important of all, it will make it possible for others to share in a profoundly moving experience, and to examine with me the meaning and value of science for the man who does it and for society. Nothing in science has any value to society if it is not communicated, and scientists are beginning to learn their social obligations.

II

How the Study Got Under Way
with the Biologists

I HAD THE money, I had the problem, and of course the general outline of procedure had long since been worked out. This is, however, a very different matter from getting the details sufficiently in hand to make a real start.

First I needed a place to work. Office space at that time was at a premium in New York, and rents high. I could find nothing suitable that the study could afford and decided to use one of the rooms of my own apartment as an office. Since I had enough standard office equipment scattered about this was just a matter of reshuffling the furniture. An office at home has more disadvantages than advantages, on the whole, and later in the study I was able to shift to a small office in a professional building. I got hold of a part-time secretary and of a young psychologist who could do some of the routine test scoring for me. There is always a lot of routine in any scientific study, and I have never found it satisfactory to have someone else do all of it. In this instance, although my assistant scored all of the tests in the subsidiary study, I also scored all of them. Later on, when there was statistical work to be done, the bulk of this was done by another assistant, but I always do some of it (which I always have redone by someone else in the interests of accuracy) in order to get the feel of the material. It is impossible, for me at least, to get this just from looking over the statistical results.

I had already decided that I would spend the first year working with biologists. There were several reasons for this, and I feel it was a fortunate choice. I did not want to start with the social scientists since I am one. It seemed best to work with them last, when I should have gained specific experience with other groups, and would presumably be more sensitive to aspects I might otherwise take for granted. So tho choice lay between biologists and physical scientists. It was an easy one to make. I had never had a course in physics in my life, and was totally ignorant of practically everything about it. On the other hand, my husband is a biologist (as well as a geologist). I had done some work with him, and read quite a little in the field, enough to be sure that I could follow all but the most technical work of my subjects without too great difficulty.

Deciding upon what subjects to use for any particular study is what is known as selecting a sample, and we speak of sampling techniques, the sampling problem and so on. For most work the ideal is a "random" sample, that is a selection of subjects which will not be biassed in respect to any of the variables under consideration. "Random" sounds as though it meant that you just went out and gathered in anyone you could get your hands on. In order to be sure that your sample is unbiassed, however, it has to be selected with the greatest care. This is an extremely important matter, because your conclusions apply only to the group of which your sample is a representative.

For example, suppose that you would like to know what 12-year olds do on a particular test. It is obviously impossible to test every 12-year old in the world. First you must decide just which 12-year olds you want to know about. This could work out in an infinite number of ways, but it might be that you wanted to know about 12-year old boys in the United States. It still is not practical to test all of the 12-year old boys in the United States, so you must test only some of them, that is, you must select a sample. If your sample is well selected you can infer from the results what all 12-year old boys in this country will do on this test. We know, however, that social and educational factors do affect test scores to some extent, so your sample must include representatives of all racial stocks and all social classes, and from all parts of the country, and these subgroups should occur in your sample in about the same proportions that they do in the total population. Obviously the sampling problem becomes easier the more sharply defined is the group you are interested in. A study of 12-year old boys of white ancestry born in this country and now living in a particular city would be quite easy to contrive.

My problem was somewhat different, and in fact I had two samples to select. First I wanted, not a random selection of competent research men in all fields of biology, but the *best* men in each field. Best men are not numerous enough that I could select a random sample from them. The reason for the choice of the best men was the assumption that if there were particular factors in the lives or personalities of men which were related to their choice of vocation, these factors should appear and possibly would appear most clearly in the men who had been most successful at the vocation. Since a great deal of work was to be done with each man I could not possibly have more than a few subjects in each field, and perhaps not more than 20 altogether in each major classification. When, then, I came to compare biological and physical and social scientists, groups of 20 would be very small. In order to make any broad inferences I needed to know how closely the eminent men in each field resembled the majority of scientists in that field. In order to find this out it was necessary to do a subsidiary study of many more scientists than the one selected for intensive individual study.

For the subsidiary study I decided to use one test and to give it to as many university faculty members as I could. It was unlikely that I could persuade them to take an intelligence test, even anonymously. This is a matter that academic people are extremely sensitive about. In any case I was more interested in personality than in intellectual variables, since we know a good deal more about the level and distribution of the latter, even in such special groups as college faculties. Therefore I decided upon the Rorschach Method since it is possible to use it as a group test, that is to administer it to a large number of people at one time. The test given in this way is not strictly comparable to the test when it is given individually but it is similar enough and this seemed the best approach available. This subsidiary study is described separately in Chapter 15, and the sampling problem with regard to it can best be discussed there.

The problem at this stage, then, was to define the field which I wished to study as sharply as possible and then to identify the most eminent research scientists in it. I decided first to limit the study of biologists to men whose work was basic research into normal life processes. This meant that I would not include those who were studying pathology, that is disease processes as such, or technology, that is the application of scientific knowledge to particular practical problems. Inclusion of biochemists in the group was admittedly debatable. So much of modern biology, though, is interwoven with or somehow stems from biochemical investigations that I decided to include some biochemists. Those finally selected have made major contributions to understanding basic life processes.

Some other limitations were necessary. We know that sex and age do affect personality and test performances. It was easy enough to eliminate variation due to sex by studying only men. There is only one woman biologist in this country who rates with the men studied so this does not distort any conclusions as applied to eminent biologists generally. It was not necessary to set a lower age limit, since the criterion that the subjects should have attained eminence would automatically do that. (The youngest man in the total group of scientists was 31.) I set a top age limit of 61. Within that range it was probable that variation in age would not affect most of the test results. (This assumption was later checked by correlational methods, as will be discussed in later chapters.) In order to keep the cultural variables as limited as possible, only men born in the United States were to be included. It also seemed desirable to omit men who had attained eminence in research earlier but who were no longer actively engaged in research,—such as men who had gone into full time administration. This is because it seems probable that the scientist who leaves research to become a college president or other administrator, differs in some personality factors from the man who remains in research, and again I was anxious to eliminate as many variables as possible. One further limitation was necessary in this group because of the personal nature of many of the data to be gathered,—I did not include my husband or any of his closest associates.

The next step was to compile a list of the active research workers in the field as defined. This was simplified by beginning with those who had already received some recognition of their work. Membership in the National Academy of Sciences is dependent upon election by peers, and indicates at the very least a high reputation for scientific work among men in a position to judge it. The same is true of the American Philosophical Society. It was a relatively easy matter to compile a list of men in these two societies whose work lay in the field of normal life processes, and then to eliminate those who were foreign born or over the age limit, or who were no longer doing research. (Lists of members are published by the societies. The other information is obtainable usually from such standard reference books as *Who's Who,* or *American Men of Science.*) To this list were added the names of other men (chiefly younger) who had been singled out by invitations to important conferences, but who had not been elected to either of these societies. (Election to these societies may take some years and the average age of the members is high. It is also true that for various reasons a man of great eminence may not be elected to either society, hence it would not do to rely solely on their membership lists.)

In order to select the most eminent men from the list thus obtained and to add others who might have been omitted, I needed the help of men, themselves in the field, who would be in a position to make the necessary evaluations which I could not do. It was also necessary to have men who represented the various branches of biology, so that every man on the list would be judged by someone competent in his field. I was extremely fortunate to be able to enlist the assistance of the following men: Dr. Detlev Bronk, physiologist and biochemist, now president of Johns Hopkins University; Dr. Hans Clarke, professor of biochemistry at the College of Physicians and Surgeons; Dr. L. C. Dunn, professor of zoology at Columbia University; Dr. W. J. Robbins, Director of the New York Botanical Gardens; Dr. J. R. Schramm, Director of the Morris Arboretum, University of Pennsylvania and Dr. G. G. Simpson, Chairman of the Department of Geology and Paleontology of the American Museum of Natural History, and professor of zoology at Columbia University. These judges rated each man on the list whose work they knew on a 3-point scale. They also added the names of others who they felt should be included. This sort of rating is not an easy job. For example, if one is rating a man on the excellence of his contribution to science, how does one balance a man who has made one or a very few major contributions against a man who has produced many not quite so major ones?

When each judge's list had been returned, I ranked the men on the basis of the combined ratings. Here, too, there were some tricky problems. Not every man received the same number of ratings, since a rater whose field was biochemistry naturally did not know much of the work of men in vertebrate zoology, for example. It seemed wise to exclude men whose work was known to only one rater (since it probably did not have as broad value as the work of others) and men who received the lowest rating from any rater, although there is one exception to the latter principle. (It seemed clear from his comments that one judge was personally biassed in respect to a man who was given top rating by several others.) It was necessary, also, to see that the final list included men from all branches, but as it happened no further adjustment was necessary for this.

After all of these ratings and rankings I had a list of 23 men. The problem then was to secure their cooperation. I had been acquainted with one of the men for some years and I had recently met one of the others, but the rest were all unknown to me personally although I had heard something of the work of many of them. It was easy to approach these two personally, and secure their help. About this time, while still worrying about how best to approach the others I had occasion to visit another city where one of the men lived. I arranged an appointment by phone, at which I explained the project to him. He expressed great interest and agreed to cooperate, but somehow we could never arrange times, and it seemed evident that he was really very resistant to the whole idea so I dropped it. But this decided me to begin with a letter for the others which would give them time to think it over. I wrote first to the men living at a distance, since visits to them would take longer to arrange if I were to fit them all into one trip. Later I wrote to the nearer ones, except for one whom I happened to meet. This was the letter, varied only as to the time suggested for the interviews:

Dear Dr. X:

The United States Public Health Service has given me a grant to make a psychological study of biologists. The purpose of the grant study is to learn something of the relations between personality and vocation in this particular field. Following other work with adults, I have for some years been doing research on personality and vocation, and have published studies on individual artists, and of a group of paleontologists (for example, J. Educ. and Psychol. Measurements 6:401–410, 1946; J. Consult. Psychol. X:317–327, 1946 and Trans. N.Y. Acad. of Sciences, Series II, 9:257–267, 1947).

I am making two approaches to the problem, the first an intensive clinical study of a few of the most eminent research biologists and the second a brief study, by a group test, of a representative sample of biologists. I should like to be able to include you in the group of eminent research men to be studied. You will want to know at once how this group for individual study has been selected. I listed the members of the National Academy of Sciences and of the American Philosophical Society in the various biological groups, omitting men who are now over 61, foreign-born or no longer doing research for various reasons. This list I submitted to my advisory committee,—all of whom were in the group themselves,—and they indicated the men in this list and suggested others whose work seemed to them to be really creative research bearing upon basic normal life processes. These are the men whom I am asking to participate in the study.

This study is one of basic research in a field in which almost no formal research has been done. My primary interest is in tracing any relationships that may exist between personality structure and the choice of the profession and the manner in which it is pursued. The results will have broad implications for assessment of the social role of scientists, as well as direct bearing on the very difficult problem of vocational selection. Beyond the vocational implications, clinical studies of normal, successful men are practically lacking, and seriously needed for better understanding of the innumerable studies of unsuccessful men.

For this purpose I will need several interviews with you, covering very generally your life history and how you happened to take up your profession. I will want to go over your publications and then to hear from you something about your reasons for doing the particular work you have done. I should like to administer several personality and other tests. For all of this, I should need about 6 hours of your time, distributed any way you like over several days or weeks. I know that you are very busy, but I feel that you will appreciate the importance of these problems to your profession and to society, and will want to cooperate.

I will not use any names in publication or elsewhere; these data will be handled only by me and will be kept entirely confidential. The results will be cast in anonymous form and will be submitted to you for approval before publication.

I have, of course, to arrange a schedule which will take me from here to the West Coast and back and meet with the convenience of a number of men on the way. Would some time around the middle of February be convenient for you?

If you have a bibliography made up, may I have a copy shortly? I should like to read as much of your work as I can before seeing you. It is possible to make up a sort of one from the Biological Abstracts, but these are not necessarily complete and in any case, indexes for the last several years are still unobtainable.

I do hope that I shall have the privilege of seeing you.

Sincerely yours,

You will note in the letter reproduced above that I stated that I would not use any names in publication or otherwise. To be asked to serve as a subject in this study was, in fact, a very high honor. On the other hand I was asking for a good deal of personal data, and I was asking the men to take a series of tests which are very revealing. Even in a group so high as this, someone has to have the lowest score and it would be most unkind and entirely unnecessary to give any indication of who this is. But there are even more cogent reasons. In some instances, such factors as divorce of parents, dissension with parents, particularly limited backgrounds and so on, come out as important elements in the over-all picture. Clearly these are matters about which any subject would be sensitive. There is, finally, the general professional training which not only makes it unethical but it simply goes against the grain to reveal anything about anyone that has been learned professionally, however undamaging or even flattering it may be. Even my records are kept under code designations, not by names. This, then, is why I present these data without names. That the men themselves for the most part make no bones about it is quite a different matter. (In fact, in the life histories, which have all been approved by the subjects before publication, there are many which are identifiable to close friends of the subject, although probably not to others. All of the test results, however, and the more personal details have been dissociated from the life histories so that they can not be related to the individual subjects.)

A very few subjects have felt sensitive about being in the group and once or twice it has happened that a wife or secretary had talked out of turn. This is, however, the most highly selected group of scientists in the United States and an incident shows that this is not unappreciated. I was gleefully informed by one subject that he had learned the identity of another. It seemed that at a bridge game the preceding summer the wife of one man was remarking with some complacency that her husband was one of those selected for this study. The wife of another was present and she quickly claimed the same honor for her husband.

After these letters were sent out I waited in considerable trepidation for the replies. If I could not get the subjects the study was a bust. The answers came in, some quickly, some slowly, and finally four men had not replied at all. Of those that did reply to this letter two declined to participate, one on the quite reasonable ground that he was about to go abroad and would not have time to fit it in beforehand, and the other without explanation. Two of the others wanted further information which I gave them. One of these then agreed, but the other still could not make up his mind and wrote, "I tried to pass the buck on the decision to my secretary who refused to make up her mind. We then decided to leave the matter entirely to chance and she flipped a coin. The fates decree that I must try to cooperate with you, and if you want to come any time after the first of the month we will see what can be done." The rest consented with varying degrees of enthusiasm. The extremes are represented in the following quotations: "Yes, I'm willing to be a guinea-pig. I must admit to a lack of enthusiasm, but after asking the advice of (a neighbor psychologist) I'm convinced that I should do my part. (Several others later admitted to having consulted a psychological colleague.) "I am deeply honored to be included in your list for study, and I shall do what I can to help."

I sent a follow-up letter to the other four, which as it happened crossed favorable replies from two of them. This time I mentioned the text-book in zoology (*Quantitative Zoology*) which I had written with my husband and received prompt and cordial replies. "Why didn't you tell me you were the Dr. Roe of Simpson and Roe?" I am not quite sure whether it was the fact that I was co-author of a text which all of them knew that established me as scientifically respectable (it could hardly have any bearing on my competence as a psychologist), or whether their acquaintance with my husband by reputation if not personally seemed a sufficient guarantee that I would not do anything too outrageous. In any case they consented to be interviewed.

This group of subjects included three in anatomy and physiology, five botanists (including plant physiologists and cytologists), four botanical geneticists, four zoological geneticists and four in biochemistry and bacteriology. That there were no zoologists other than genetical is a reflection of the fact that the most active research in the field at that time was genetical.

By now, some months had elapsed, although I had also succeeded in getting the subsidiary study under way. I was fortunate in being able to see my old acquaintance first, which was a great help in many ways. The first few interviews were not as efficiently conducted as the later ones, and I am sure that the second and third years' work was somewhat better. On the other hand, when I was travelling for the second year I revisited most of the men I had seen the first and was able to fill in the gaps nicely at that time.

Another factor in the superiority of the later interviews was my own increased skill at shorthand. I had taught myself shorthand before doing the work with artists. I had dabbled in it beforehand, but really worked at it for that purpose, since it is much, much better to get test responses on both the Rorschach and the Thematic Apperception Test down verbatim. An interview recorded verbatim is also very much more useful than one for which only notes have been jotted down. I had given some consideration to the use of a recording device, but had decided against it on several grounds. I was afraid it would make the subjects self-conscious for one thing, whereas all of them were accustomed to dictating to a secretary. I also felt that lugging it around the country would be awkward; I had to take a typewriter and brief case as well as a small suitcase as it was. Then, too, in the actual interview situation there were advantages to my busying myself with a shorthand record. To some extent it would depersonalize me, an advantage to both of us. I wanted to ask as few questions as possible, and then to let the subject carry on in the order that occurred to him. When the inevitable pauses occurred, I could let them go on while still recording twirls. The result is a rather subtle form of pressure that usually will induce the subject to continue without further direction, and this is often very useful psychologically. There is the still further factor that this system contributed materially to my own ease in the interview,—I had not to worry about disturb ing the subject by watching him too much, and I didn't have to smoke all the time.

This system had, however, the very great disadvantage to me that I then had to transcribe all of the notes which was a considerable burden. I felt this should always be done before a succeeding interview, in the case of the life history data so that I could note what had been missed, and in the case of the tests so that I could do a quick scoring and interpreting job on the spot. This was desirable, because I tried always to see the subject again after I had given and scored the tests. They were usually interested in the results naturally. A succeeding visit also gave me an opportunity to check some of the interpretations and very often to get material of a sort it would be difficult to get so quickly in any other way. For example a life history might have been given that contained no indication of any particular parental conflicts, but some of the test data might indicate very clearly that there had been such conflicts. This might or might not be pertinent to the study, in the end, but it gave a chance to check whether the test indications were correct and to get further personal material. It takes at least as long to transcribe the notes of an interview as the interview itself takes, and this became extremely burdensome. The last year I carried a Soundscriber with me, and every night read the interview record onto it mailing the discs to my secretary. I still had to transcribe the tests for immediate use, but this was much less than I had been doing. This change had the further advantage that I was not too exhausted to record general observations which I had not done earlier in as much detail as was desirable.

This, then, was the general procedure with each subject as it was soon worked out. If possible, before seeing him I had read at least some of his work, enough to know what his major interests were and how he tackled them. Usually, however, I had not been able to do a thorough job of this and it was also much quicker for me to borrow from him a set of his papers (most scientists keep a set bound up for their own use) and to go through them during the period of my visit. I arranged for three different interviews whenever I could and I consider this the ideal system, although two can be used or even one. I always got life-history first, however many interviews there were. In getting this I usually explained that I wanted to know the following things: something about his family so I could get the general social and economic background; his brothers and sisters; his early schooling and his preferences; what he did outside of school and during vacations; everything he could tell me that had any bearing on his choice of a vocation; how he had learned of his profession, what other things he had considered and why he had finally decided as he had; his college and professional training; his

professional history since college (I usually had this in outline from the reference books); what he now did in his leisure
time; his religious interests; and finally in what terms he
thought. Then I let him go ahead and give this to me as fully
as he wished, and asked questions only when necessary to
clarify a point or to remind him of something he had omitted.
With some, of course, much more questioning was needed
than with others.

The item about the terms in which he thought had not been
part of the original plan. After I had talked to a number of
the biologists, and discussed their work with them, it suddenly dawned on me that there seemed to be a difference between the way in which they were thinking and the way in
which I customarily thought. I had no reason to think that
I was particularly atypical for my own group, and it occurred
to me that this difference might have some significance for
the problem. From then on I made a particular point of this
and went back to the men I had seen earlier with further questions about it. In the end this gave some of the most interesting results in the whole study. These are discussed in Chapter XI. It is not too unusual, in starting out in a new field, to
come across something which you had never thought of looking for, and this is a good example. This whole study is still
in the earliest phase of a scientific investigation, an observational one. At this stage it is important not to have fixed ideas
of what you are going to see, and to look at the situation as
comprehensively as you can, or you may miss something or
distort what you do see. After such a study as this has been
done you can set up hypotheses about why you have seen
what you did and then test these out one at a time on other
groups. What you have observed, if you have done the work

properly, stands as a correct account of what took place in the group you studied. Any causal relationships are only inferences at this stage, however, and it is these that must be checked. What you want to know finally is not just what happened, but also why it happened, that is, which factors led to which results.

By the time we had finished such an interview as this, the subject was usually pretty much at ease with me, and I with him. As a rule I then gave him the Rorschach and the Thematic Apperception Test. The biologists usually did these rapidly but you never know in advance how long they will take and have to allow plenty of time.

After this the next step was usually to run over his bibliography with him, and discuss many of the items. By then I would have read a good deal of his work. I wanted always to find out why he had done that particular study when he did, what led up to it and how it fitted into his overall program. These were invariably fascinating discussions, and my subjects were very patient in explaining things to me. It is these discussions on which I shall rely heavily in working on the problem of the creative process itself. I have never yet had time to work over them thoroughly.

Finally I gave the test I call the Verbal-Spatial-Mathematical Test. I call it this, or preferably VSM for short, because it has three sections, one a test of verbal ability, one a spatial test and one a test of mathematical reasoning. This is also described in a later chapter. It is an intelligence test,—one which had to be especially constructed in order to get one difficult enough for this group.

I also obtained handwriting samples when I could. I tried to get them not only as of the present, but from every five or ten years, and from as early an age as possible. I have not done anything with these as yet. The scientific analysis of personality from handwriting is a technique that is being worked out at the present time. When it has been sufficiently developed it will certainly be an extremely useful technique for biographical studies, since it is practically the only one conceivable which will permit analysis of the personality not only as it is at the time of the analysis but as it was at any time in the past. This should give us extremely important clues to changes in particular aspects of personality not just with age, but also with particular incidents in the life history.

In our long discussions it was often quite natural for me to mention my husband and I found that this made a considerable difference socially. I am quite sure, for example, that I was more often invited to my subjects' homes because I am Mrs. Simpson than because I am Dr. Roe. I thoroughly enjoyed such visits, and of course seeing a man in his own home often gave me much more insight into the kind of person he was. My subjects would have been kind, in any case, but this was different and it had its amusing aspects. At the first university I visited, for example, where I had several subjects to see, I had made reservations at the local hotel. When he learned who my husband is, one of my subjects immediately exclaimed, "Why didn't you let us know? We would have arranged for you to stay at the campus club."

I was particularly fortunate in this first stop on my long trip. I was still very jittery about the whole thing, but all of the men were most pleasant and cooperative. Two of them took me home to dinner and asked other pleasant people, and I had another fine evening with the local psychologists. It was not all hard work, by any means. When, the day I was to leave, I suddenly came down with a current local virus, the wife of one of my subjects rallied promptly to my aid, and I spent several extra days there, mostly getting better acquainted with all of them. By then, one of them felt free enough with me to tell me that he had been quite disturbed when I first showed up in his laboratory,—I looked so unlike what he was accustomed to seeing there. As his wife had also remarked that she had seen me on the campus the first morning, and had obviously immediately identified me, it seemed clear that my appearance was a little too urban, or something. I took the hint gratefully and never again walked in on a subject for the first time wearing the sort of ridiculous hat and costume jewellery that I normally wear. (The trouble is that ridiculous hats are often the easiest kind to pack!)

When I got to my next stop it turned out that no time had been lost as one of the men had forgotten I was due several days earlier and had gone up to the country for a few days. I prudently kept my psychological deductions to myself. Although he did his best and I did mine, he was never really happy about any of it. The other subject there was very interested, extremely friendly and cooperative, and I still remember a very pleasant dinner in his home.

Here there was an interlude for personal affairs. I was headed for the west coast, and could go through Albuquerque. We had purchased some property 90 miles from there, and were planning to build a home the following summer (where this is now being written). There were still many details to be attended to, particularly as the house was to be started before we would be able to get out, so I arranged to stop over for a day. One of our neighbors had adjusted his regular visit to town to meet me and drive me up and I planned to catch the bus back that night and go on next morning. The neighbor had warned me that it was bitter cold that February, with temperatures around 30 below, so I had ransacked the stores at my last stop for some woolen underwear. I got off the train wearing the woolies, a heavy suit, a wool blouse and my winter coat, plus a warm scarf, only to find there had been a sudden change and it was then 60 above! Even without my coat the drive up was somewhat warm, but I was glad of all my things later, for I had remembered the bus schedule incorrectly, and only got back to Albuquerque by catching a ride with a milk truck in the middle of the night!

My next two stops were not less pleasant. My subjects and their families were friendly and at one I was able to stay with a colleague. This was delightful but had its trying side. She not only knew my subjects but is a very acute person, so that I had to be extremely careful of what I said. It was the more difficult as several of the tests I obtained there had particularly fascinating technical aspects, and it was hard to keep away from shop talk under those circumstances. I did succeed, I think; professional training is very strong in this respect. (It seems also to have been strong enough to overcome my usual habit of full discussion of everything with my husband who knows no more of the study than anyone else, except inevitably, the names of most of my subjects.)

I have, of course, discussed some of the tests with some of my colleagues, but only as tests, and only under circumstances which precluded any possibility of the subject's being identified. In several instances I felt I needed another opinion on some point. For example, one of the Rorschachs I obtained in the course of this study had some characteristics which have been noted as frequently present in cases of organic brain damage. I am not an expert in such diagnosis from the Rorschach and described it to a colleague who is. If he had felt that the existence of brain disease was actually possible I should have had to do something about it. Deciding what to do had kept me awake more than one night, but fortunately for my peace of mind the expert felt that the indications that had worried me were overbalanced by other facets of the test, and that it was unlikely that organic damage was present. This is only one example of the sort of unanticipated and very difficult problem that can arise in any clinical psychological research.

Both the rest of this trip and subsequent trips to other subjects were of the same order. I counted myself extremely fortunate to have begun with so helpful and friendly a group, and I was greatly relieved that there had been no serious difficulties. Such procedures as these interviews and these tests can, and quite frequently do, arouse a good deal of anxiety in the subject, or they may bring back unpleasant and difficult memories. Of course these things did happen, and then one must to some extent at least, act more or less in the role of therapist. Under these circumstances this is an extremely subtle matter. I felt strongly that the overall effect should be of some value to the subject himself, and in most cases I think it was. It often served to help him clarify things in his own mind to which he had not given much thought, but which were of interest or importance to him. It also often brought back pleasant memories, and the process of trying to state why he had followed his profession was frequently illuminating to him.

On the whole, so far as I was able to observe it, I think the subjects derived more pleasure than pain from the procedure, even those who had a particularly difficult time. There was one who burst into tears at one point, to the great surprise of both of us. There was another who nerved himself up to telling me first about a very unhappy and disorganized episode in his past and then went on feeling, I thought, much relieved to get that off his mind. There was another who finally asked me to give the test results to his psychiatrist, thus disturbing me considerably until I had talked to the psychiatrist, for I did not wish to interrupt the course of his treatment, and I would have proceeded with even more caution had I known previously he was under treatment. (I had no doubts as to why, but many intelligent people do not think of consulting a psychiatrist or psychologist over their problems.) And there was the man who kept his appointment with me immediately after he had learned that one of his children had not long to live. There was, he said, nothing at all he could do for a few hours and it would help him to have to work at something, if I thought it would not distort the results. I have the profoundest admiration for his courage and control. I have a profound admiration for all of my subjects. In fact, a critic of the technical monograph on the biologists complained that it was clear that I was "emotionally involved" with my subjects. This is what a psychologist says when he means that you like someone enough to feel strongly with him. He was quite right.

III

The Year's Work with Physical Scientists

I FELT VERY much at a loss when the time came to start work with the physical scientists. My knowledge of their subject was practically limited to recent journalistic accounts of one special area, nuclear physics. I had never felt really sure that an unplugged piece of electrical apparatus would not electrocute me, out of sheer meanness, or that radio was really anything more than mass hypnotism! Clearly something drastic had to be done about this. I dug in for a couple of months with several dozen books, but alas, some of them are still incomprehensible to me. I began with the army textbooks, which are about high school level. I got through these all right, and did not do too badly with an elementary college text, although the lack of experimental experience was a serious drawback. Thanks to the atom bomb there are a number of well-written expositions of nuclear theory at a comprehensible level, and I read those. Such books are practically lacking in other branches of physics, though, and then I was stuck. I felt somewhat better about it, however, when one of my subjects assured me that he could not understand Bridgman on thermodynamics either. I did succeed in getting a good enough idea of what went on that I could follow the major lines of what most of my subjects were doing, if not how they did it. One result was, however, that I spent less time reading their work, and much less in discussing it than I had with the biologists. This does not introduce any serious deficiencies into the study of the relationship between choice of physics as a profession and personality structure but it does lessen the amount of data I have on the creative process in physics.

The difficulty of the sampling problem was also greater for me with this group. Richard Tolman, whose wife is one of my most valued colleagues, had agreed to help me with this and his untimely death left me without an advisor His colleague, Dr. Paul Epstein, very kindly came to my rescue at this point and devoted a great deal of time and effort to this aspect of the study, and also helped materially in getting some of the subjects to cooperate.

Again, the subjects were to be males, American born, and still engaged in research. Dr. Epstein constructed a list of men doing basic research in the field of physics proper, and in astrophysics, geophysics, physical chemistry and theoretical engineering. It included both men whose primary interest is theory and men who are primarily experimentalists, but it did not include any men whose primary work is on practical problems.

The raters for this group were Dr. Epstein, who is a theoretical physicist at California Institute of Technology; Dr. Walter S. Adams, an astronomer at the Mt. Wilson Observatory; Dr. Hugh Dryden, chief physicist of the National Bureau of Standards; Dr. Beno Gutenberg, geophysicist of the California Institute of Technology; Dr. W. V. Houston, now president of Rice Institute and a mathematical physicist; Dr. George Kistiakowsky, physical chemist at Harvard; and a nuclear physicist who prefers to remain unnamed. These men rated the work of the 69 men on the list submitted to them and added four others. Three on the original list were reported by raters to have gone into full-time administration and had to be dropped on that account. Following this procedure, 30 men were selected. These included men from all of the groups named above, and both theorists and experimentalists. Because of the number of refusals and because the categorization under which several of the men were rated had later to be changed, the final selection is not completely balanced. (For example, three of the physical chemists are theorists and only one an experimentalist.) There is only one astrophysicist and one geophysicist in the final group, but there are relatively very few in these groups as compared to all physicists.

I had known one of these men slightly but I was correct in thinking he would not recognize my professional name and he was quite surprised when I appeared. I had sent him the same letter I did the others. This was substantially the same as that quoted in the last chapter, but I added that I had completed a year's work with biologists. Of the 30 men who were asked, only 22 became subjects. This is a higher refusal rate (27%) than for either of the other two groups but there are some special circumstances which must be taken into account. The most important of these is that physicists have, in actual fact, less time and are under much greater pressure than either of the other groups. This is true even of those men, very few indeed, who do not still have government projects or consultative relations with government departments.

The two who refused on a plea of inadequate time were probably not misstating the case. Even those who accepted had less time to give than the biologists, and it not infrequently happened that I was able to secure only one interview, instead of the two or three which I preferred. In such instances, I was usually given a whole afternoon during which, with the reduction of time required for discussion of their work, it was possible to cover the major data, except for the VSM test. In these instances the test was left with the subject. It is easily self-administered, but of the 4 so left, only 2 were filled out and returned to me.

I do not know why the other 6 refused. It is entirely conceivable that they just did not wish to be bothered and I must admit that I can understand this very well. It has been suggested, too, that physicists have become hypersensitive as a result of the numerous inquisitions to which many of them have been subjected, and this may well have been a factor. There were five who did not answer either of two letters, but three of these I was able to reach by phone and they readily consented to take part in the study. All explained that they just do not answer letters!

There may have been another factor which I did not suspect at the time. I learned from several who did become subjects that they had suspected something in the nature of a disguised Kinsey interview and were greatly relieved that this was not what I was after. The possibility of such a misconception had never occurred to me. Kinsey is not a psychologist, but his book on the sex habits of the American male had just appeared and it is true that psychologists had entered into the discussion of it particularly vociferously.

There are still other differences among these groups than

the number of refusals in response to the invitation to participate in the study. Quite a number of the physicists took the precaution of checking up on me and some of these checks were very thorough. Several of them asked Dr. Epstein about the study. In a number of instances they called an acquaintance who was a psychologist. In one instance the subject requested me to have two of the psychologists on his university's faculty call him. I did so and later learned that he had put them through quite an intense inquisition about me. They did well by me, as all my colleagues have, and he did become a subject. Still another discussed the matter with a colleague to whom I had also written and they agreed mutually that the latter would be the better subject; but this was after I had again been checked on. The man who had been asked about me had reported, among other things, that I had done some extensive studies of foster children and that my husband is a paleontologist. This led to one of my favorite cracks. When he saw me this physicist asked whether I had picked physicists for study because I suspected they were the most childish or the most fossilized of scientists.

Obviously if your subject has assured himself that you are professionally respectable the interviewing is much easier, and certainly this is a procedure I would myself follow in similar circumstances. It was interesting, however, that more physicists than biologists did follow it. In such small groups this difference could occur purely by chance. It could, however, be interpreted either as reflecting an entirely rational hypersensitiveness, as already mentioned (although I saw little evidence of this among the men who did see me) or as reflecting a difference in character structure. I am inclined to think the latter is a factor for reasons which will be brought out later.

With this group I followed much the same program of travel as I had with the biologists. As before, the visits were uniformly pleasant and stimulating although I had fewer social contacts with the physicists. On the whole curiosity about what psychologists are up to seemed a stronger motive in this group than among the biologists. There were a number, however, particularly those who had had some administrative problems during the war, who felt that eventual solution of many personnel problems waited on such basic research as this.

As always there were a number of unexpected incidents. There was, for example, the man with whom I had arranged an interview with great difficulty, who turned out to have been born abroad. This is the only such error. It occurred because this datum was not in the standard reference books and none of my raters knew it. He had been brought here at such an early age, however, that I decided to include him anyway.

One of the men I knew to hold an administrative post but since he had continued doing research he had been kept in the list. I was, however, somewhat disconcerted to find that another was about to switch over to full-time administration, but since he had not at the time of the interview he was also included. The loss of research men to administration has a number of interesting aspects. It is definitely my impression that more physicists give up research, or teaching and research for full-time administrative posts than scientists of any other group. If this is so it raises some interesting questions. Physicists do tend to become eminent younger and perhaps this is a factor in their being considered for administrative posts. That they do attain eminence younger, on the average, than other groups of scientists is at least partly due to the nature of the field. As one physicist put it: "I think it's true that scientists reach distinction earlier, more or less in the ratio of the relative importance of intellectual effort to the experimental work. In pure mathematics you can have a smart idea and become a great man. In biology you essentially have to make experiments, you can't speculate about what an animal is doing and this is hard and takes a long time." I suspect, however, that another factor may also be involved in both the early achievement of eminence and the frequent shifts to administration. That is the rate of maturation and of decline of some special abilities, such as are represented in the space test discussed later. Space perception and possibly numerical ability, reach a peak at earlier years and decline faster than do verbal abilities. On the other hand the rate of decline may be less rapid among those with greatest amounts of the ability. (This is an example of the sort of speculation in psychology that would be very easy to check, as far as technical problems go. There might be some difficulty in getting a sample, although in my experience this problem is not a serious one. The only serious problem would be financing such a study. The actual cost would not be small, but it would be much less than the results would be worth.)

There were a number of particularly delightful incidents. There is, for example, the physicist who introduced me to one of my favorite "laws," which he described as "Murphy's law or the fourth law of thermodynamics" (actually there were only three the last I heard) which states: "If anything can go wrong it will."

Another physicist quite rightly thought it would interest me to see a cyclotron, and so took me around. I can report seeing enormous piles of concrete, and a control room full of switches, lights, and all sorts of elegant gadgets, but that is as far as I can describe a cyclotron for you.

One series of interviews was particularly moving. One of the men had replied cordially and favorably to my request for an interview, but added that he was then convalescent and that he did not know whether this would affect the tests or not. I had assumed from the tone of his letter that he was convalescing from some ordinary illness and hence was deeply shocked when I called on him to find him dependent on an oxygen mask. It was apparent, however, that he welcomed the opportunity for some distraction and indeed he gave me a great deal of life history material and discussed his work with great fluency and ease. It is important, too, that in the verbatim record of this discussion, except for some pauses, there is no evidence whatever that he was in any serious way hampered intellectually by his illness. I thought it would be worth while trying the Rorschach and the Thematic Apperception Test but I would have omitted the intelligence test since it is timed and the physical set-up made his situation completely incomparable to the others. He knew of the tests, however, and wished to try them all, so we did but I have not used any of them. The effect of the illness upon his working capacities showed very clearly in the test situation. He could, of course, do only a little at a time, but the difficulty went much further than that. The Thematic Apperception

Test and the Rorschach were too difficult as problems in organization of new material, and this was in sharp contrast to the good organization in the narrative of his life. The Rorschach also gave a picture that suggested some organic brain damage, but was unlike any other I had seen or read of. After his death a few months later, I communicated with his physician and learned that he had suffered severely from cerebral anoxia (insufficiency of oxygen in the brain tissues) during the last few months of his life, and it was apparently the early stages of this that the Rorschach had picked up.

A number of these men are seriously concerned with the problem of student selection, especially with the greatly increased demands for men trained in the fields of physical science. With one of them I had a most interesting discussion of a problem in graduate training which has developed in the last few years. This is the problem of the married student whose wife expects him home every evening at five or five-thirty. This is, after all, a reasonable expectation for most wives in our culture, as my subject admitted, but from his point of view it showed complete lack of understanding of the situation. (He also admitted that so far as he knew no one had explained the situation to the wives.) It is impossible, in his opinion, to turn out a properly trained and qualified Ph.D. in physics with only four years of graduate work if this is on a 40-hour week. It requires much more time and study than that. Formerly, he said, the students thought of practically nothing but their work and the lights were burning in the laboratories every night until late hours. Now the laboratories are dark at night.

This is, I think, not basically a nostalgia for the "good old days," and there are some quite crucial considerations. Formerly there was a considerable, if often somewhat harsh weeding out of students which was at least partly and often almost entirely on the basis of motivation. By motivation I mean eagerness to enter the profession in spite of obstacles, and absorption in it. It was so difficult to get through graduate school, economically and otherwise, that it was chiefly those to whom it mattered terribly who made the grade. There is probably no more important factor in the achievement of this group of scientists than the depth of their absorption in their work. Now there are so many fellowships and other forms of financial aid for students who qualify on the basis of intelligence or undergraduate records that there is practically no selection on the basis of motivation. There is also some good evidence that high motivation is best maintained in the face of some, but not too many or too difficult obstacles. It is a serious question in my mind as to whether the amount of aid now available not only may reduce the effectiveness of selection of graduate students, but also may act to cut down eagerness in the students selected. If so, I think that this is a serious deterrent to the development of first class research scientists.

IV

The Year's Work with Social Scientists

Now I HAD reached the point of working with my own group, with the social scientists. I have called the psychologists and anthropologists social scientists in order to have a convenient single term for both together, although this is not a very good term. It is not a very good term because it does not apply very well to experimental and comparative psychologists or for that matter to physical anthropologists, and it would apply equally well to sociologists whom I did not include (largely because of limitations of time). What anthropologists and psychologists have in common is primarily that they are interested, in somewhat different ways, in people, in how they differ, and in how and why they behave as they do. Anthropology which means the study of man would be a good term for both groups, but it has already become so firmly established as the name for only one group that it is not possible now to use it in the larger sense. Anthropologists and psychologists have as much or more in common as the various subgroups of biologists. My three major groups, then, the social, the biological and the physical scientists, are definable as representing scientists interested in man, in all living

50

things and in the non-living aspects of our world. There are, in each instance, other specific groups who would belong in any of these (for example, geologists in the last group) but it has not been possible to study every group, and here too there had to be a sort of sampling. With the completion of the study of the social scientists, I felt I had enough data to begin extensive comparisons.

As before the same limitations of age, sex, nationality and present research activity were imposed. For neither the psychologists nor the anthropologists were the membership lists of the National Academy of Sciences or of the American Philosophical Society useful. Representation of both groups is small which is logical enough for the anthropologists since this is relatively a very small group. In addition, psychologists in either society who meet the study's limitations are experimentalists. There are no clinical or social psychologists, for example, in either organization.

For the psychologists the preliminary list was made up by me in consultation, separately, with Dr. E. G. Boring and Dr. David Shakow. We simply went over the membership list of the American Psychological Association and put down everyone we knew to be actively engaged in research and otherwise qualified. This preliminary list was then rated, in the usual fashion, by Dr. Boring, of Harvard University, Dr. E. R. Hilgard of Stanford University, Dr. D. B. Lindsley of Northwestern University, Dr. Jean W. MacFarlane of the University of California, Dr. David Shakow of the University of Illinois and Dr. L. M. Terman of Stanford University. Dr. Boring is a historian of psychology and psychologists, Dr. Hilgard an experimentalist with a particular interest in learning theory, Dr. Lindsley a physiological psychologist, Dr. Mac-

Farlane and Dr. Shakow are clinical psychologists and Dr. Terman has been closely identified with testing and with the study of genius.

As before, the ratings of these judges were combined, and the men who ranked at the top were selected, with some adjustment so as to include representatives of different sorts of psychology. In one instance, when a choice had to be made between three men of equal rank and working in about the same field, I tossed a coin. Three psychologists declined to take part,—one of these later spontaneously reconsidered and is included. A number of psychologists who did agree to serve as subjects did so with some statement of reluctance. Such a statement came most often from those with at least some clinical training. Whether this meant that they were actually more reluctant or just freer to admit it I do not know. If they were more reluctant it was probably because the clinicians had a very good idea of what was involved and must have felt that the tables were being turned on them.

The original list of anthropologists was made with the help of Dr. Robert Lowie, and with his advice a list of 82 names was arrived at. For this the raters were Dr. A. V. Kidder, archeologist of the Peabody Museum at Harvard University; Dr. A. L. Kroeber, cultural anthropologist at Columbia University; Dr. R. H. Lowie, ethnologist at the University of California; Dr. A. H. Schultz, primatologist and physical anthropologist, then at Johns Hopkins University; and Dr. C. F. Voegelin, specialist in linguistics at Indiana University. Of the 13 men selected on the basis of their ratings, one was out of the country during the period of the study, 3 refused, and another consented but was finally unable to take part because of illness. The 8 in the final study include, in broad classification, 2 physical anthropologists, 2 archeologists and 4 cultural anthropologists.

One of the anthropologists, before agreeing to become a subject, asked a question which I had wondered at not having been asked before. This was the question of what would happen to the records eventually, and it is a very good question, indeed. The answer was that they were covered by a codicil to my will, leaving them to the library of the American Philosophical Society which had agreed to accept them under certain conditions. The conditions are that the record of each man is to remain sealed until 10 years after his death and that it is then to be used at the discretion of the librarian of the society. I had chosen this society not only because many of my subjects are members but because its library specializes in the history of science in America and because it is an organization which has lasted over two hundred years and gives every evidence of remaining in existence for a long time to come. Lacking such an arrangement I would have had to direct that the records be destroyed, which would seem a great pity since there is such fine biographical material in them. All of my subjects have agreed to this disposition of the data. Incidentally my lawyers in drawing up the codicil were very troubled about who could sue whom if anyone did not like it (which apparently is something lawyers like to be sure about) since I would be dead and my heirs could hardly be blamed for what I had done!

This year's work was of extraordinary interest to me, because since my years as a graduate student I had read relatively little in psychological fields other than my own, and I found it interesting and exciting to get up to date in learning theory and physiological psychology and animal experimenta-

53

tion and all the rest of it.

One interview had to be interrupted while my subject performed an operation on an experimental animal. This was a very interesting business. In other places I had a chance to see laboratory apparatus that was new to me and to see many techniques in action that I had only read about before.

The work of the anthropologists was equally fascinating. I had already read some in this field; I have lived in odd places, and known little-known people, such as the Kamarakotos of the Venezuelan wilds, and I spend much of my time now in the American southwest where the various Indian cultures are intensely interesting.

By then, too, my techniques were smoother. (So much so that one psychologist wrote, after seeing my report, that he was amazed that I had been able to get so much information in so casual a manner. What could have been more flattering?) My shorthand was also much better. This was fortunate, because most in this last group are rather given to talking and my notes go on for pages and pages and pages. I was, after all, a colleague and they were naturally somewhat freer in talking to me on that account. They were also much more sophisticated with respect to the sort of thing I was interested in and had more appreciation of the possible relevance of many details of their life history which sometimes did not occur to the others. Of course we could also spend a good deal more time in discussion of the research of all of these men, since my own understanding of the work and its background was much greater than it was for the other two groups.

Naturally also on the trips to see these groups I had a good deal of social activity. In fact, I had even more fun than before, on the whole, even if I never did get enough sleep. I often revisited subjects from other groups. In several universities I was asked to give seminars on this study and the ensuing discussions were often of great value to me.

The question of giving tests to testers naturally arose. The intelligence test I was using was one put together for the purpose so that there was no problem of knowledge of that particular test. There is very likely a sort of test-wiseness that psychologists may be expected to have, and this must be taken into account. It did not seem likely that this would result in a significant raising of the scores on so difficult a test (and in any case I had no exact check on this) and I have not made any formal correction for this. It happened that, although a few of these subjects had some first-hand knowledge of the other tests, none was expert in their use, and after some consideration there seemed no cogent reason for omitting them.

It was often possible, with these psychologically sophisticated subjects, to go into greater detail in the test interpretations although they, too, are sensitive. This was of particular interest and value for the extra checks it afforded on the use of the tests, even though the subject's own opinion, however informed, is a very hazardous criterion. (There is no psychological test of any value which is completely self-interpreting, and every clinician knows that he must constantly be on guard lest he develop personal biases, or slip into a sort of rule-of-thumb interpretation for tests he uses regularly.)

V

Social Description of the 64 Subjects

THAT, THEN, is how and why the study came about and how it was done. It is time now to see what was found out. Perhaps it will be easier to get the feel of these scientists as people, if I summarize first a few of the more obvious things about them, age, marital status, children, and what they do with their leisure time. We can then go on in the next chapter to inquire into their backgrounds and the sort of experiences they had in growing up.

As has been stated, the top age limit was set at 61, and a lower age limit was set, in effect, by the requirement that all the subjects should have achieved eminence. Within these limits, age was ignored in making the final selection. There are some differences among the groups and these are shown in Table 1. Some reasons for these differences were suggested in Chapter III.

All of these men were married at the time of the study, and most of them have children. The average age of first marriage is about the same for all of the subgroups and is rather late, 27 years. The youngest marriage was at 19, the latest at 35. That they married late, on the whole, in comparison with the

Table 1

Age of the Scientists at Time of Study

	RANGE *	AVERAGE
Biologists	38–58	51.2
Physical Scientists	31–56	44.7
Social Scientists	35–60	47.7

* The range shows the age limits of the group, that is, the youngest and the oldest.

general custom in our country is partly due to economic and educational factors. Not many of them could afford to marry before they had finished their training, but it is also true that many of them were too intent upon their work. This is in sharp contrast to the situation today, when a much higher proportion of graduate students are married than was the case twenty years ago.

At the time the study was made, three of the biologists and one of the physicists had been divorced, and nine of the social scientists (several of whom have been divorced more than once). All have since remarried. Since the study was made, another of the biologists and another of the physicists have been divorced and have not remarried. The differences between the groups are very striking. Expressed in per cent the divorces for each group (at the time of the study) are: physical scientists 5%, biologists 15% and social scientists 41%. When one remembers that several of the social scientists have been divorced more than once, the differences are even more striking, and are clearly too great to be due to chance. This situation is consistent with some of the other findings which will be discussed later. Here I will just point out that social scientists are very much concerned with personal relations, with people as people, and that the other groups are not. Part of the reason, I think, why they have more marital problems than the other groups is that their marriages, in one sense, matter more to them, hence they may demand more of their wives and perhaps their demands seem excessive to the wives. It is also possible that they more easily become involved with other persons and that this puts an additional strain on the marriages.

Only 10 of these subjects have no children. The others have from 1 to 4 children each, with an average for the total group of 2. (At least 3 children have been born since the study was made, but I think there will not be many others because of the ages of my subjects.) Their families are smaller than their parents' families which is generally characteristic of the population now. Their parents had from 1 to 7 children with an average of 3.

I asked them how they spent their leisure time. Of course some of them said "What leisure time?" This is not just a gag since many of these men work nights, Sundays and holidays and a number do not take any regular vacations. For the most part these long hours are a matter of choice. They say, for example, "My real recreation is doing what I want to do, my work." "I have no recreation. My work is my life." "There is nothing I'd rather do. In fact my boy says I am paid for playing. He's right. In other words if I had an income I'd do just what I'm doing now. I'm one of the people that has found what he wanted to do. At night when you can't sleep you think about your problems. You work at it on holidays and Sundays. It's fun. Research is fun. By and large it's a very pleasant existence."

Of course there are other people than scientists who work

most of the time, and some of these do it because they would rather be working than doing anything else. But I think rather few people are so fortunate as to be able to earn their living by doing what they most want to do. There are so many jobs that leave parts of you unused, that give little scope for individual planning and initiative, and it is these needs that hobbies and recreations should meet, as well as a real need for relaxation.

Most of these subjects, though, do have a few moments at least apart from their work. The bulk of this time is spent with their families. Some of them will even go so far as to take their children to the movies, an occupation that otherwise most of them avoid since the practically universal opinion among this group is that, with very rare exceptions, movies are worthless and boring. They are not alone in this opinion. A report on leisure interests of 2340 professional men (not including scientists) states that only 32% at ages 40 to 49 years attend movies. I suspect that the reason for the general dislike of movies is that nothing is left to the observer but just passive observation, for one thing, and for another that so many of the movie situations are totally irrelevant to any part of life that has meaning for scientists. The stereotyped emotional situations that seem to be the fate of movie stars have little or no counterpart in normal living, but what is more important is that they also have no attraction for these men. The ones who are not interested particularly in human relations in real life are not likely to be interested in them in the movies or television; the ones who are interested prefer them in fact and not as the cultural stereotype may have them.

The amount and kind of reading they do is extremely varied. Some read nothing but professional literature, and others read a good deal, including detective and adventure stories, science fiction, classical literature and science in fields other than their own. Reading is a favorite recreation of professional men.

About a quarter of the group have some sports interests (about the same proportion as other professional men) although these are less actively pursued than they were earlier by most. Fishing, hunting, climbing, sailing, swimming and tennis were particularly mentioned. It is noteworthy that these are largely individualistic pursuits, i.e. team play is lacking.

Only four of them have played any active part in political or civic organizations although a few others are occasionally gotten into such activities by their wives, quite a number of whom work in the League of Women Voters. Most of them do vote, but there are a few who make a point of not doing so. I would judge that their political views ranged from rather rightist to very leftist, with the bulk of them definitely liberal.

In the leisure activities so far mentioned the three main groups do not differ. There are some differences among them in other activities. The physical scientists and the psychologists have more active musical interests; more of them have played and still play instruments than is true of the others. The most striking difference, however, is in their social interests. There is a small number of the physical scientists who are very socially inclined, and who spend quite a lot of time partying, but most of them and practically all of the biologists dislike social occasions, except perhaps for very small gatherings of close friends. In general they avoid social occasions as much as possible and rather resent it if forced by professional duties or their wives into spending time this way. This is in marked contrast to the social scientists among whom a great deal of voluntary social activity is common.

There are a few among the total group who have highly specialized interests, almost hobbies. For example, several are interested in particular sorts of historical research. One composes music. Several write novels or poetry.

None of my subjects spontaneously mentioned church activities as important to him. I usually made a point of inquiring about religious interests although I do not have definite information on this from 10 of the subjects. The results are very striking. Nothing was known of the religious background of these men when they were selected, hence this was not in any way a factor in their selection. In the total group of 64 scientists, 5 came from Jewish homes and all but one of the rest had Protestant backgrounds. The parents of the one exception were strong "free-thinkers." There were none from Catholic families. Other studies have shown that Catholics rarely become research scientists. This is not surprising when it is considered that the research scientist must have freedom of inquiry. This freedom must be as much within himself as within any social framework; he must be able to accept what he finds without having to force it into any already given scheme. The Protestant churches in which most of these men were brought up have varying degrees of insistence upon the authority of the church over its members' interpretations of life. Some of them, certainly, are as restrictive as the Catholic but they seem not to have so strong a hold.

Although the intensity of the religious interests of the parents varied a good deal it was customary to send the children to Sunday School. All of the large Protestant denominations are represented among the parents of my subjects and a number of smaller ones, Mormons, Quakers, United Brethren, etc.

Even with this background the fact is that now only 3 of these men are seriously active in any church. A few others attend upon occasion, or even give some financial support to a church which they do not attend, but they are not personally concerned over religious matters, at least within any institutional framework. All of the others have long since dismissed religion as any guide to them, and church plays no part in their lives, personally, although they may send or permit their children to attend Sunday School. A few are militantly atheistic, but most are just not interested.

The statements of a few of my subjects on this point are given below. I have included statements representing the extremes of attitude, as well as the most common ones among this group.

"I find more wisdom concerning mind, values, conduct and therapy in Christian doctrine than I do in our beloved but still puerile science of psychology."

"I'm not very much of a church goer. I'm a great believer in church but don't get around to it. I always went as a boy. Mother was a very good church woman. I'm not attached to any church but I go occasionally. I usually go where my wife wants to go. I had no sort of religious crisis; if anything science strengthened me on that. It was so clear that one doesn't understand things. There is this business of checking back on nature and finding you were wrong in your ideas. I've had so many clear cut ideas and set up the apparatus and it came out a different way, it makes you pretty humble when you think about it."

There were several who went through periods of severe conflict. One was brought up in a very orthodox Methodist household, with no dancing, no theatergoing, and so on. His

father's success in business meant that in private school he
was with quite a different group. "When I was 16 what to do
about dancing caused me agonies." He solved the problem
of his class dance by taking the minister's daughter, and they
did not dance but just walked around and talked. When he
was 17 he began to waver in his religious beliefs. "Father
lived until I was 48 and mother until almost then. I was anxious not to give offense to them, so I postponed admitting I
was an agnostic, but I've had no conviction since I was 18.
About 20 I realized I couldn't honestly say the Apostles'
Creed."

"I can remember one thing when I was shifting from one
church to another. (This was when he was in college and
going through a period of worry about religion.) The Con-
gregational minister wanted me to join and this caused me
a great deal of distress and I felt very guilty. Why was I different because I couldn't believe it? After I discovered most
scientists didn't believe it it was all right. I kept going to
church when I was in college because I went with a couple
of girls. I think this made me unhappy because I felt deeply
over it for several years."

"My parents were very religious and belong to a group
where the religion is quite emotional. I couldn't understand
what was going on and couldn't appreciate it and I was be-
wildered and sometimes frightened. I think at the age of 7
I went through what was supposed to be conversion but I
didn't know what it was all about. Perhaps it was because
my older brother did but I felt let down. I didn't really feel it,
it was just sort of cadging. Church attendance was required
and Sunday was very strictly observed in our family. At college I attended church services and Sunday School and was

a leader in Christian Endeavor for a year or so. That was quite a liberal church. I had no serious crisis when I left. I had never taken the fundamentalist point of view. I had always questioned it and tried to analyze religion as such. I don't think the family took my defection very seriously."

"I am pretty completely agnostic as to what is behind the scene of things. I'm willing to take my chances on its being a fundamentalist or Catholic god. If such a being should turn out to be in charge I think the only self-respecting thing is to be damned but I suppose I have a vague humanistic attitude towards the situation."

"Religion was one thing we were saved. Mother went to church when she was asked to sing and usually went once more for every time she was asked. Father would go occasionally with her because it was the proper thing to do. And I went to Sunday School and had the best necking of my life, but I don't think I've ever been in a church since."

"Father and mother were both free thinkers. Father was somewhat aggressively so and enjoyed nothing more than insulting ministers. The result was that I grew up as a free thinker. I was sent to Sunday School until I was 9 or 10 and then I read Paine and Ingersoll at 12 or 13. This had a certain prestige value. I would do things like this. When we were out shooting I would point my gun at the sky and say 'I hope this bullet hits God.' The free thinking set me aside in a way. I think since childhood I have had a conviction that the majority is always wrong."

These quotations are representative of the attitudes of the total group. It seems the general rule that whatever the religious interests of the parents were the children have tended to non-religion. Practically all of this group did receive at least some religious instruction as children. (Even the son of free-thinkers was sent to Sunday School.) Only a small minority now have any connection with or any interest in established churches, and to only three of them is religion of any great importance.

VI

The Family Backgrounds

LET US TURN now to consideration of the sort of background these men came from. We might ask first in what part of the country they were born. It will be remembered that one was born abroad. The other 63 came from 25 different states, 20 from the east, 25 from the midwest, 14 from the west and 4 from the south. The distribution of their birthplaces by states is given in Table 2. There is some tendency for the largest number to come from the states with the highest populations. This is not consistently the case of course, but the group is too small for any meaningful geographical analysis. Some came from farm homes, some from very small towns, some from large towns, and some from big cities. These geographical factors, then, seem not to be very significant.

The economic level of the families from which these scientists came is extremely varied. Three of the men experienced serious deprivation as children and others came from relatively poor homes. Some of them came from very well-to-do families, but there are none from enormously wealthy families. Most of them come from what would be called upper middle class homes. This range is the same for all of the sub-

Table 2

Birthplace by States of 63 Eminent Scientists

EASTERN STATES	20	MIDWESTERN STATES	25
Connecticut	1	Illinois	6
Maryland	2	Indiana	4
Massachusetts	6	Iowa	4
New York	8	Kansas	2
Pennsylvania	3	Minnesota	1
		Missouri	1
WESTERN STATES	14	Nebraska	4
California	5	Ohio	1
Colorado	2	South Dakota	1
Oklahoma	1	Wisconsin	1
Oregon	2		
Utah	2	SOUTHERN STATES	4
Washington	2	Louisiana	1
		South Carolina	1
		Texas	1
		West Virginia	1

groups but a higher proportion of the theoretical physicists and of the anthropologists come from well-to-do families.

Occupation of the father is probably the best single indicator of the social and economic status of the family. Data on this are given in Table 3. If the father shifted occupations the one he followed during most of the childhood of the subject is the one tabulated. Since there are interesting differences among the various subgroups they are given separately, except the biologists who do not fit neatly into a smaller grouping.

Table 3 shows that in respect to the occupations of their fathers this group is very unlike the population of the United

Table 3

Occupations of the Fathers of 64 Eminent Scientists

	BIOLO-GISTS	EXPERI-MENTAL PHYSI-CISTS	THEO-RETICAL PHYSI-CISTS	PSY-CHOLO-GISTS	ANTHRO-POLO-GISTS	TOTAL
PROFESSIONS	9	5	10	7	3	34
Research Science	0	1	0	0	0	1
Physician incl. optometrist	0	2	1	2	0	5
Lawyer	0	0	1	1	3	5
Engineer	0	0	3	2	0	5
Clergyman	2	0	1	0	0	3
Editor	2	0	0	0	0	2
Teaching: College	4	0	3	2	0	9
Elem. or H.S.	0	2	0	0	0	2
School Superin-tendent	1	0	0	0	0	1
Pharmacist	0	0	1	0	0	1
BUSINESS	8	1	2	4	5	20
Own business	4	0	2	2	4	12
Clerk, agent, salesman	4	1	0	2	1	8
FARMER	2	4	0	2	0	8
SKILLED LABOR	1	0	0	1	0	2
TOTALS	20	10	12	14	8	64
% Professional	45	50	84	50	38	53

States in general. According to the census reports for 1910 only 3% of the gainfully employed men in the country were professional men. In my group of eminent scientists the figure ranges from 38% for fathers of the anthropologists to 84% for the fathers of theoretical physicists and is 53% for the total group. Furthermore there are no sons of unskilled laborers among them and only 2 sons of skilled workmen.

In 1921, Cattell and Brimhall made a study of the families of scientists. They found that 51% of the fathers of 66 leading scientists were professional men. That two studies, made for quite different reasons and 30 years apart should agree so well on this figure makes it reasonably certain that in this country about half of our leading scientists are the sons of men in professions. It follows, of course, that most of the fathers of scientists (and it is also true of the mothers of my group) had had a better education than that of the populace at large.

It is not enough to find out just that this relationship occurs. The question is, why does it occur? It is quite clear from the table that it is not that the sons follow their father's professions. Only one of them has, exactly. The sons of those who were teaching college are themselves college teachers but only one is working in the same field his father worked in. The relationship is much less specific than that. There is some correlation between occupation and intelligence, and level of intelligence is believed to be at least in part hereditary. There is no doubt about the superior intelligence of this group of scientists and one must consider whether this relationship results just from the fact that the sons of professional men have a somewhat better chance of being of superior intelligence. But this explanation would not account for the sons of men in relatively humble occupations who became scientists unless one goes on to assume a higher level of intelligence in the father than would appear from his activities, or that the son inherited his intelligence from his mother. Either of these is possible but unprovable on the present data, and in any case the inheritance of intelligence is far from a simple matter.

It is my opinion that the effective factor is a rather different one and I base this on the situation, not only in the homes of most of the professional men, but in many (not all) of the others. This is that for one reason or another learning was valued for its own sake. The social and economic advantages associated with it were not scorned, but they were not the important factor. The interests of many of these men took an intellectual form at quite an early age. This would not be possible if they were not in contact of some sort with such interests and if these did not have value for them. This can be true even in homes where it is not taken for granted that the sons will go to college.

There are some differences among the groups in the general family expectation of sending the sons to college. As nearly as I can tell, college was taken for granted for 65% of the biologists, 60% of the physical scientists (but it was very different for the subgroups, being 84% for the theorists and 30% for the experimentalists) and for 82% of the social scientists. These differences are associated with other differences in background. In general the sons of professional men and of well-to-do business men were just automatically expected to go to college.

There is another very interesting point of resemblance between this study and that of Cattell and Brimhall, referred to above. This is in the numbers of scientists who are the first-born in the family. The data for this group are given in Table 4, where it is seen that 39 of these 64 scientists were the first-born. Of these, 15 are only children. This is many more first-born than would be expected by chance, and the excess is just about the same as that reported in the earlier study. For this calculation, Cattell and Brimhall used 855 scientists of moderate but not the highest standing.

Table 4

Position in Family of 64 Eminent Scientists

POSITION *	NUMBER OF SCIENTISTS
1	39 **
2	13
3	3
4	3
5	2
6	2
7	2

* Under position, 1 means first-born, 2 second-born and so on.
** 15 of these are only children.

We may say, then, that being first-born increases the chances of becoming a scientist. But why does it? Before theorizing we might look at the ones who are not first-born. There are 25 of these. Of them, 5 are the oldest sons, and 2 who were second-born are effectively the oldest during their childhoods since the older children died at birth and at age 2. Complete data on 3 of the others are lacking; they are all youngest children, one of 7, one of 5 and one of 3; in the case of the first two of these I do know that the next older brother was much older than the subject but I do not know exactly how much older. For the remaining 15, the average number of years between the subject and his next older brother was 5. It would seem, then, that most of those who are not first-born are either oldest sons, or substantially younger than their next older brothers. There are only six of whom this is not true and these will be discussed individually later.

It would seem from many aspects of these results, that the development of personal independence to a high degree is an extremely important factor in the production of a research scientist. It is possible that the situation of being oldest son is one which carries with it in many instances a larger amount of independence than other family positions do. On the other hand, 15 of these were only children, and a few were clearly overprotected, which would at least to some extent work against developing a feeling of independence. They might, and sometimes did, however, get considerable indulgence in the matter of pursuing their own interests which was of benefit to them.

If one considers the common family situation in which a younger child is frequently baffled because he cannot quite do the things a next older child can do, and must often give up on this account, or at least feel inadequate, the frequent occurrence in this group of considerable distance between the subject and the next older brother becomes of great significance. When there are enough years between them, competition is not acute in the same way and it is much more easily taken for granted that the older brother is stronger or can do things the younger brother cannot. If particular encouragement is not present, at least a particular sort of discouragement is absent.

What happened with the subjects who are just a little

younger than a brother? (The presence of a slightly older sister is of much less importance to a boy. He is rarely in direct competition with her in ways that matter to him apart from such general competition as exists for the parents' affections, and even here the sex difference may make this much less explicit.) One biologist is just one year younger than his only own brother but both of these boys were very early on their own since they did not get along with the stepmother. Furthermore their interests were so different as materially to reduce any competitiveness.

Another biologist had a brother about 2 years older, with whom he did compete for such things as school grades. The boys were frequently separated for various reasons, but I think there is no doubt in this instance that the competitive situation served to make the younger brother put forth more effort instead of to discourage him.

One of the physicists, the youngest of 7 children, had a brother only a year and a half older than he. I do not know enough of this family to be sure what the situation was, but apparently all of the children were pretty independent.

One of the psychologists was the second of 4 boys, with his next older brother only 2 years older. This family was one in which the boys were very close and supportive but it is also true that this next older brother was ill and bedridden a good deal during their childhood which would reduce direct competition in many ways. Apparently it did not function to increase competitiveness for the attention of the parents, but there were other affectionate adults very much in the picture.

One of the psychologists whose next older brother was 4 years older seems to have felt the situation more strongly than the others, and said that he had always felt he could not compete with his brothers in any way and speaks now of deep-seated feelings of inferiority.

It would appear, then, that there are certain factors in the family situation into which a child is born which are somehow associated with his becoming a successful scientist. At this point, we do not know whether the primary association is with becoming a scientist or becoming successful, but further research will clarify this. One of these factors is a home in which learning is valued for its own sake. This is most likely related primarily to choice of profession. We find such homes particularly often when either or both parents are better educated than the average and the father is a professional man.

Another factor seems to be position in the family,—being the oldest child, the oldest son, or at least a number of years younger than the next older brother. One may speculate that this is more closely associated with success than with choice of science but one cannot be sure. It is also possible that if the older brother is athletically inclined, competition would turn the younger to other pursuits so it might be in some instances effective in the choice of profession.

VII

Growing Up

"IN A LARGE family I learned to read before I went to school. I was in the first grade for two days and when the teacher found out I could read she took me to the principal. They tried me out and I could do better than the third grade, but they thought that wouldn't do and so they put me in the second. Reading was pretty nearly my life in the early school days, and writing. I was practically always the top of my grade in most everything. Then about the time I went to high school we moved to a farm and father wanted the farm run scientifically. That began to fascinate me and I read Morrison's *Feeds and Feeding*, a great big technical book. I realize now that what intrigued me was the scientific part. We would plant a plot of this and a plot of that and keep records of it."

"In grammar school I just loved the subject of grammar, that just was the apple of my eye. Arithmetic I couldn't do. The only way I ever passed was that father worked the problems and I learned them. In seventh grade we had a teacher who taught agriculture and it was very good and my first contact with anything at all like science and I liked it fine. It just kept me on the front edge of my chair all the time."

"I just can't remember that anything in grade school was important. I had very poor teachers except for one very good teacher who taught me how to study. I think I passed out with a grade of 76 as I remember it. In high school I liked math and Latin, but wasn't very fond of physics. I had an ambition to be a high school teacher of Latin. I think only in college did I really get interested in science."

"I can't remember much about grade school except that I got reasonably decent grades right along and that I was fairly interested in science and math. I had a friend in seventh or eighth grade who was the son of a druggist and we got a chemistry set between us and played around with it and almost blew up the house. We spent our spare time memorizing the table of elements. I never got along in languages; I couldn't see any sense in memorizing grammar. In history I read so much I had many more facts than the rest, whether they were right or not. I didn't like physics in high school but I took an extra course in chemistry. There were only 4 students, and the teacher let us do pretty much what we wanted. This convinced me I wanted to be a chemist."

"In high school I was enthralled by the classics. Then I went into a chemistry class by requirement and it happened that the teacher and I hit it off and so I got to be sort of an unofficial lab assistant and I got interested in chemistry. Then in some way I got hold of a copy of E. B. Wilson's *The Cell* and I read that and the *Origin of Species*. But then I had a teacher of zoology and one time he had one of his classes get a human head and boil it up and make a skull of it. That revolted me and ever since then I've had a dislike of any laboratory contact with animal materials."

"I had no interest in any scientific thing as a kid. I was interested in athletics chiefly and if I had any vocational ideas it probably involved becoming a coach. I liked chemistry and math in high school and college and did them well and easily. Latin and modern languages were very difficult."

These are a few quotations from remarks about early schooling and school interests. They have been selected more or less at random, and they illustrate the very wide range of degree of interest in school subjects. Can you guess what kind of scientist each of these men finally became? The first became a psychologist, the second an experimental physicist, the third a physical chemist, the fourth a theoretical physicist, the fifth a botanist and the last a biochemist.

By far the majority of these men went to public schools. This is true of all of the subgroups except the anthropologists, all but two of whom had at least part of their schooling in private schools.

It is also of interest to inquire into what they did outside of school hours, when they were free to choose their own activities. In the life histories which follow, you will read about some of these activities in detail but here I will summarize the reports for the whole group.

Among the twenty biologists, only 10 showed any interest in natural history as such when they were children. For 5 of these this was rather vague, but the other 5 were intensely interested in special aspects,—they hunted flowers and birds, they kept records of their finds and they spent long hours working at them. The 10 who did not show special natural history interests did not have any very special interests before high school days. They played around with the other children, but they do not remember anything outstanding. When they got to high school, several of these became strongly interested in chemistry. A few had courses in agriculture which interested them from the scientific angle. For most of them this was not only their first but also their only contact with science up to that time. Courses in biology were practically never given in high school at that time, apparently, and physics was less commonly available than chemistry. At the present time there seems to be a better variety of science courses available. Whether or not they are better taught I do not know.

Children tend on the whole to take school for granted, but more of these children had a positive liking for it than not. The school subjects which particularly appealed to them varied. Only two or three of the biologists were particularly interested in literature although a number were great readers. More of them were interested in chemistry, or any other science they happened to take, or in mathematics.

Among the physical scientists, both theorists and experimentalists, there are many reports of early intense preoccupation with gadgets, with radio, with Meccano sets, and so on. This was quite rare in biologists and social scientists. Both groups of physicists showed considerable early preference for mathematical and scientific subjects in school. The theorists, however, were strikingly omnivorous readers. Except for one who was more interested in athletics for a time, they usually made some such comment as that they read everything they could get their hands on. A few of them soon concentrated on science but a number were interested in biography and history. Two of them remarked that they got their first interest in science from reading science fiction.

Among the social scientists the commonest early interest was in literature and the classics. A few had some natural history interests, but most, like the theoretical physicists, spent a large part of their time reading. There are a few who were early interested in mathematical or scientific courses, but not many. Of course there are practically no activities a boy can indulge in that are definitely related to what psychologists and anthropologists do professionally. A rare exception would be an early opportunity to become acquainted with Indian communities or artifacts.

When did these men make their decision to become the sort of scientist they did? Over half of them made it during their junior or senior years in college, but some made the decision in early childhood and some as late as the second year of graduate school. It tended to be later for the social scientists than for the others. This is, at least in part, because of the fact that few of them were in a position to find out about psychology or anthropology very early. The social scientists also seem to have been somewhat later in worrying over the necessity for making a vocational choice. There is a further factor that a number of them originally decided upon a literary career, and did not discard this until they found literature an unsatisfactory technique for interpreting human behavior. They were not satisfied largely because neither literary creation or literary criticism was sufficiently precise, that is, sufficiently scientific. A few others were first interested in a social service career, and spent some time on this but they later discovered the possibilities of psychology or anthropology and found them more satisfying.

You may be interested in how old these men were when they finished various parts of their formal training. The average ages at which they completed college and graduate school are shown in Table 5. The physicists, particularly the theorists, tend to be more accelerated than the others in their school careers, but it should be remembered that fewer of them had to work their way through.

Table 5

Average Age at Receiving College Degrees

	B.A.	PH.D., SC.D., M.D. (EARNED)
Biologists	21.8	26.0
Physical Scientists	20.9	24.6
Social Scientists	21.8	26.8

More important than when they decided to become a particular kind of scientist is why they decided to. It was not just a matter of early interests. As has been said, they had a variety of these, and some were and some were not related to what they finally became.

From fiddling with gadgets of various sorts to becoming a physicist is not a very great step, once the discovery is made that one can make a profession of this. To go on to become a theoretical physicist requires some contact with the specific field, since it is not so obvious. This is also true of anthropology and psychology, and indeed in biology, too, it is frequently the case that the possibility of a research career is not learned of in early school years. In the stories of the social scientists and of the biologists it becomes very clear that it is the discovery that a boy can himself do research that is more important than any other factor in his final decision to become a scientist. I think this is also important in physics but there the discovery comes so gradually as not to be noticed as such. In the other sciences it often came as a revelation of unique moment, and many of these men know just when they found this out. A few quotations will illustrate this and it is further documented in some of the life histories which follow.

"I had no course in biology until my senior year in college. It was a small college and the teacher was about the first one on the faculty with a Ph.D. It was about my first contact with the idea that not everything was known, my first contact with research. In that course I think my final decision was really taken. It was mainly that I wanted to do something in the way of research though I didn't know just what, but working out something new."

"The transfer came in my senior year in college. I was taken over to the university principally because the chemistry professor was interested in locating a teaching assistantship for another student. I had my eyes opened then and saw students doing research work. It was just one afternoon, but the research interested me, the idea of using chemistry to find out new things. I must have been very excited. While I was there I applied for a teaching assistantship which I didn't get, but I went on and started course work."

"One of the professors took a group of us and thought if we wanted to learn about things, the way to do it was to do research. My senior year I carried through some research. That really sent me, that was the thing that trapped me. After that there was no getting out."

Once any of these men had actually carried through some research, even if of no great moment, there has never been any turning back. A few of them feel that they would be equally happy in some other field of science but only one has ever seriously wanted to do anything else. This is a Nobel prize winner who has always wanted to be a farmer but could not make a living at it.

That a research experience is so often decisive is a matter

of considerable importance for educational practice. This is true not only because the demand for scientists is so much greater than the supply, but also because this is a very satisfying manner of life for all who have pursued it wholeheartedly. The discovery of the possibility of finding out things for oneself has most often come about through the experience of working on problems individually, rather than of reading what others have done. Sometimes this is the result of careful preparation on the part of the teacher; sometimes it happens because the teacher is more interested in other things and leaves his students to work on their own.

How many other boys would have turned to science if they had ever discovered the possibility that they, personally, could become scientists? Much of our educational system seems designed to discourage any attempt at finding things out for oneself, but makes learning things others have found out, or think they have, the major goal. It is certainly true that it is easier to teach a set of "facts" than it is to encourage an inquiring mind (which most children have to start with) to make its own discoveries. Once a student has learned that he can find things out for himself, though, bad pedagogy is probably only an irritant. I feel strongly that this problem goes much further than just the matter of developing more scientists and will say more about it later. More widely conceived it means developing more citizens who can think for themselves and this is the basis for greater social advances, and greater democracy. It is also a basis for much greater happiness for individuals.

There are other, less obvious, factors that seem to have played some part in the choice of vocation these men made. These are things that happened to them in their homes or

elsewhere, that served as important molders of their personalities and characters and that seem somehow to be related
to their liking for what they are now doing, and the energy
they have put into doing it.

It is very difficult to be sure of such things as the effect of
different childhood situations such as broken homes, illness,
and family discipline. This is because we have so little information on what generally or most often happens to children when they are growing up, so that it is often hard to say
whether the things that occurred in this group would or
would not occur in any group taken at random. Wherever
such information is available I shall include it. (Such a situation as position in the family is quite easy to check because
we can calculate what the probabilities are of being first- or
second-born and so on.) When a situation occurs very frequently, or when there are marked differences in the frequency of occurrence among these groups, it is reasonable
to ask if there can be any relevance, and to look for psychological meaningfulness. Then a hypothesis can be set up and
checked on other data.

Here are a few quotations about home situations.

"We had a whole series of housekeepers, that is all I can remember about growing up. Mother died when I was about 4.
I think the main effect of it may have been just coincidence
but I had a rather unusual social life. It wasn't so much lack
of opportunity but I didn't take much advantage of it. I had
a lot of friends in the neighborhood and spent time in the Boy
Scouts but not in mixed activities. I got crushes on girls but
they never knew it. I got over it in college in undergraduate
days."

"Discipline was liberal but with fairly rigid standards, but coercion was subtle. We were all good children, we never had a hand laid on us. We were indulged, we never did any chores. Mother kept us dependent on her in some ways and she could get us to do what she wanted us to. We developed a pretty strong sense of moral oughtness."

"My father is a man who values independence and self-reliance and he is anxious to see his sons have it."

"My father was exceedingly hostile where I was concerned and exceedingly unpredictable. I could never tell when he came into the room whether he was going to be nice to me or knock me down."

It is not possible to reproduce in full detail all of the life histories, although a number are given in the next few chapters. Here I will try to summarize some of the situations that seem to be important.

One of the first things that stands out is the frequency with which these subjects report the death of a parent during their childhood. The data, together with information on divorces of parents, are gathered into Table 6. The age at which the loss occurred is important. I have somewhat arbitrarily divided the group into those whose loss occurred before and after the age of 10, in part because I have some comparable figures on this basis. Also a younger boy is less likely to be put into the position of becoming the man of the family on the death of his father. Fifteen percent of the total group lost a parent by death before the age of 10, and the figures are 25%, 13% and 9% for the three major groups. In only one instance, as noted, was a satisfactory substitute provided. In some other instances the surviving parent remarried but the results for the child in terms of affection and care were not satisfactory.

Table 6

Age of Subject at Loss of Parent

	UNDER 10				OVER 10			
	BY DEATH			BY DIVORCE	BY DEATH			BY DIVORCE
	Father	Mother	Tot.		Father	Mother	Tot.	
Biologist	2	3	5	1	0	0	0	1
Phys. Scient.	2	1	3	0	2	0	2	1
Soc. Scient.	1	1 *	2	0	3	1	4	1
Total	5	5	10	1	5	1	6	3

* In this instance a satisfactory home was provided by foster parents.

The only contemporary figures I have been able to obtain were supplied by Dr. Robert Strauss. He found that in a group of 624 college students only 6.3% had lost a parent by death before the age of 10. His group would be roughly comparable to this one in terms of social and economic position. The increased expectation of life which has come about in the last 30 years would explain part of this great difference, but even so it is clear that the biologists are well beyond normal expectation in this respect.

In a group of 183 homeless men, Dr. Strauss found that 25.2% had lost a parent before the age of 10. The literature is full of statements, based upon such figures as this, about the adverse effects of broken homes. Broken homes have even been cited in themselves as a *sufficient* cause for any number of difficulties. But look at the figures for eminent biologists who have come out at the other end of a scale of social usefulness. Obviously a broken home, in itself, is not the sole cause of becoming a bum,—or a biologist for that matter. Certainly the loss of a parent is a serious extra stress in growing up, but it need not be always disrupting. It isn't the extra stress, but what you do about it that matters. For some people extra difficulties call forth extra effort.

It would be interesting to check biographies of other great men of other times. The only ones that I have happened to come across where this is easily done are in the collection of biographies in the book *Men of Mathematics*, by E. T. Bell. He records the lives of 32 mathematicians and physicists, from Archimedes to Cantor in some detail, and while he does not always give data on this point, it is definitely stated that 8 of these men (25%) lost a father or mother before the age of 10 and 10 before the age of 14.

Perhaps such a loss as this can serve to increase the degree of independence attained at a relatively early age. The attainment of independence seems to be essential to advance as a scientist. But it may also be that such a loss has other, more direct effects in individual cases. The fact that in this group this situation has occurred most often among the biologists is suggestive. A very good technique for handling difficult emotional problems is to generalize them into a less personal one and sometimes to reverse them. In a few instances in this group it would seem that the problem of facing death had been turned into a deep concern with the mechanisms of life; in such a situation the strength of the motivation shown in the pursuit of research can be explained by the depth of this concern. There are others in the group, however, for whom no such relation is suggested. In these instances what seems to have happened is that this loss helped to establish a kind of life that did not include a need for close personal relationships.

Among these scientists I think that the divorce of the parents was not of great significance. The biologist whose parents were divorced when he was 9 had a difficult time for a few years but there was a grandmother to help give him emotional support. The divorced parents of the social scientist shortly remarried and the home was constituted as before. For the two others whose parents were divorced, the divorce seemed to alleviate a difficult home situation, and in each instance the son was by then quite self-sufficient.

In all of these groups there are a number who had trouble while growing up because of some exceptional physical condition. A few were abnormally tall, a few were abnormally small, and some seemed to be unusually weak. Such situations were somewhat commoner among the social scientists and the theoretical physicists but I know of no comparative figures on this point, so I cannot estimate its significance.

There does, however, seem to be a significant incidence of early physical problems among one of the subgroups, the theoretical physicists, when both such conditions as those referred to above and serious childhood illness are considered. Only 3 of these 12 men had generally good health and normal physical development. Apparently a major result of this was an increase of a sort of social isolation which is also reported by these subjects. A number of them were out of school for long periods. That most of them were avid readers was partly intensified by these physical difficulties. Illness or other physical problems did not occur to any extent among the experimental physicists, and severe illness had low frequency among the social scientists and the biologists.

It is possible that this common physical situation may have been a factor in developing an attitude which is rather specific, so far as I can make out, to physicists. It is held by experimentalists as well as theorists, but perhaps to a lesser extent. This is the manner of thinking of the size of objects. To most people, concepts of size are directly related to their own bodies as they perceive them, or to the body image as psychologists call it. This is true even though people often do not think of their own bodies in this connection. Physicists, however, who may deal with galaxies one day and atoms another must be able to think of size in completely abstract terms. If, in growing up, your body has proved rather unsatisfactory, it may be easier to get away from the usual approach to size.

The problem of social integration or isolation may also be of some importance in these groups. Many of these men had quite specific and fairly strong feelings of personal isolation when they were children. They felt different, or apart, in some way. I suspect that these feelings are very common generally, and that almost everyone has them at one time or another. They would be significant, then, only if they persisted over long periods, were particularly strong, or had some other unusual aspect. So far as my records show they were least common in these groups among the biologists but I think that this may be because this was the first group I worked with and I have somewhat less personal information about many of them than I have about the others. Such statements as the following from the physicists seem particularly strong:

"In college I slipped back to lonely isolation."

"I have always felt like a minority member."

"I was always lonesome, the other children didn't like me, I didn't have friends, I was always out of the group. Neither the boys nor the girls liked me, I don't know why but it was always that way."

Among the social scientists this sort of statement is much commoner:

"The family was essentially self-ostracized. There was a great confidence in our complete intellectual superiority."

"We were a family that kept completely to itself."

"We had a feeling of being somebody in our small town."

"We developed forms of living which were different from those around us."

It seems quite clear that there is much oftener great closeness of family relations among the men who became social scientists than in the other groups, even allowing for differences in their reporting. There is also, with this group, the additional factor of strong feelings of superiority, either on a personal, or more frequently, on a family basis. Such feelings of apartness as they had were much more often colored with an attitude of superiority.

Furthermore the whole problem of personal relations has always been much more important to the social scientists, with the exception of a very few among the experimental psychologists. It would appear that to most of the men in the other groups other things mattered much more. This meant, among other things, that they had fewer parental conflicts, and that they more easily achieved an attitude of independence from their parents. This situation is clearly reflected in the Thematic Apperception Test results as well as in the life histories. The social scientists, on the other hand, had many more severe and open conflicts with their parents, and more hidden ones, and quite a number of them are still angry about it. This means of course that they have never quite succeeded in breaking away completely.

If one looks closely at the family situations, there appear

to be some that occur much more frequently among the social scientists. In a number of their homes the mother was a dominant figure and the father was rather ineffectual or was a man who suffered from some feelings of inadequacy. These seem to have been sensed by the children at the time, although probably not clearly. In other homes, in which the father was clearly dominant several other situations occurred. In a number of these the father was a self-made man who had achieved a good economic position by intense effort; in several such instances it was remarked that his mother (the subject's grandmother) had had strong social pretensions. In several others it was remarked that the subject's mother valued her husband's economic position for the justification it gave her for social strivings of her own. In still others social superiority was taken for granted. In short, in about three-quarters of this group, social status was of conscious importance during the childhood of the subject.

When social status is important, high value is naturally placed upon interpersonal relations. (A child will accept such valuations without thinking about them. He will accept stealing as smart if his group places high value upon such an accomplishment.) People then become invested with unique importance, and if interpersonal relations are difficult for some reason the problem then becomes one of crucial importance, and attracts much of the available emotional energy. One way of handling problems is to think of the personal problem as just one aspect or example of a generalized problem, and then to work on the general problem. This cuts down a lot of the personal painfulness. It may not be done on a conscious level at all, but we have observed it repeatedly. It is psychologically a very sound technique. It not only decreases the amount of unpleasant emotions involved but it serves as a motivation for doing a good deal of work that may be extremely useful. I remember one psychologist whose remarks about his father were strangely reminiscent. I checked, and sure enough, he had written about a group he was currently battling professionally (and I think with good reason) in almost exactly the same terms. And he was quite startled when I pointed this out to him.

One might comment that the vocational activities of the psychologist and anthropologist, particularly of the social or clinical psychologist and the cultural anthropologist, are very satisfying in a particular respect. The position of studying one's own or another society requires a certain Jovian viewpoint, which in itself has a sort of reassurance of personal superiority. It also permits this reassurance under circumstances which are not only not asocial, but definitely of social value, and hence doubly acceptable.

(If it occurs to you that there is a certain analogy here in the scientist who studies other scientists I can only say that it has occurred to me, too.)

The greater disinterest in persons shown by the biologists and physicists has other effects. Many of them were slow to develop socially and to go out with girls. This could be partly on a defensive basis, because many of them did feel shy and inadequate in these respects. But the more important thing is that it did not matter enough to most of them to do anything about it. (If it had mattered so much that they had turned to other things in real desperation, one could almost certainly pick this up in their present attitudes, or from their test performances, and I am sure that this has rarely been the case.) There is a characteristic pattern of growing up among both biologists and physical scientists. Of course there are exceptions in both groups but they are relatively few. The pattern is that of the rather shy boy, sometimes with intense special interests, usually intellectual or mechanical, who plays with one or two like-minded companions rather than with a gang, and who does not start dating until well on into college years. Even then dating may be a very secondary matter. In several instances the subject first developed a strictly platonic friendship for a fellow graduate student, and after that he was able to date some other girl in a non-platonic way.

This is in great contrast to the social scientists, most of whom became openly interested in girls much earlier. Half of them started dating in high school and dated happily and extensively from then on; only four did very little or no dating until they were through college. I have a decided impression, not only of much more dating, but also of much freer sexual activity generally.

These developmental differences are paralleled in the present attitudes shown by these men. The difference in present social interests of the groups has already been noted. Most of the biologists and even more of the physicists regard their fathers, if not with great affection, at least with very genuine respect. Relatively few of the social scientists do. Psychologists would say that it had been much easier for the physicists and perhaps to a less extent, for the biologists to "identify" with their fathers. It is currently believed that part of the normal process of growing up is for a boy to sort of feel himself to be like his father and that this helps him to become a man himself, and makes it easier for him to behave as a man is expected to behave. Certainly physics as a profession has a more "masculine" tinge by our standards than psychology has, although these social standards seem in process of considerable change at the moment. Actual or supposed inadequacy of a father, and certainly dominance of a mother can easily interfere with this process. (Perhaps this is an even more important, but subtler factor in the greater number of divorces among the social scientists than the factors that have been mentioned previously.)

VIII

Becoming a Biologist

I CANNOT TELL you the life stories of all of these men although all of them are interesting, and of course all of them are different. Perhaps if I tell you several of the stories for each group and point out how the others differ, when they do, you will get some idea of what kind of childhoods they had. I will take one of the biologists who had an early, intense interest in natural history and knew what he wanted to be from the start, one of those who rather drifted into science, and one who found out about it suddenly.

Let us call the first one Henry which is not his name. When Henry was born his family lived in a middle-sized town, and his father taught in the state college there. They lived in a big house with large grounds, and they and many others in the town had vegetable and flower gardens, and even wild flower gardens. He used to help his father with the spring planting. He has many memories of playing in the gardens and some of them go back a long way. He recalls:

"When I was 4 I was given some beans to play with, weevils got into them. I lived in an enormous Dutch house and in the spring when the frost was going out of the ground the man who came to deliver the wood left wheel tracks all over. So I planted the beans in the wheel tracks and then it rained. I may have patted down the dirt. Anyway I went away on a vacation and while I was gone the beans came up. At first no one could imagine what had happened in the front yard, and then mother remembered giving me the beans. They were all gone by the time I came home.

"When I was a child a brook ran in back of our house and a sandy beach, and I had a sandpile down there and when I was about 6, I was fascinated by the way people made geranium cuttings. I snipped off some of my mother's geranium shoots when she wasn't looking and carried them to my place by the creek and rooted them. I expected to be punished for doing it. I can still remember how astonished I was that I wasn't punished but mother was proud I had pulled off something like that."

He remembers two early scientific experiments. He had noticed that the clouds seemed to follow him around, and he wondered if they really did. He was afraid to ask because he had found out that people were apt to smile at his questions. He impressed his baby brother (still in long dresses) into service, telling him to stay quite still and to watch the cloud. Then Henry walked down to the end of the garden. But his brother reported that the cloud had not followed him at all. The other experiment was a botanical one.

"The next experiment I remember I was 8. I was always thrilled every spring with the things planted in the garden and always helped. Every year father had planted morning glories and nasturtiums. I had heard about hybridizing plants, and I decided I would hybridize nasturtiums and morning glories. I started out with typical and superb self-confidence.

I took the morning glory seeds and pushed them into the seeds of the nasturtiums. I had no doubt that they would grow. I planted them in the edge of my wild flower garden. The morning glories didn't come up but some of the nasturtiums did, but even they looked a little sick. I expected either they would be morning glory flowers on nasturtium vines, or nasturtium flowers on morning glory vines, but I wasn't sure which they would be and that was one reason why I was so fascinated."

Now Henry is a geneticist, but this is almost too pat. There are others who early were intensely interested in natural history, but whose eventual professional work did not necessarily follow in just the same field. One of them, who became a botanist, spent all his early years collecting and studying birds. The names of only a few birds were common knowledge. Then he saw a new one and described it to his sister who described it to a teacher who identified it for him. This was a tremendous experience, to know that each bird had a name and that he could find it out by giving the facts to someone who knew.

There was another who spent most of his spare time as a boy collecting eggs, carefully taking only one from each nest. Sometimes he would spend all morning watching one bird in order to find the nest. He kept a field book (he still has it) with a note of all the birds he found, and when and where he found them, and finding a new bird was one of his greatest satisfactions. He collected butterflies, too, but now he works with neither.

Henry remembers also that he was fascinated by shapes when he was a boy. He could put picture sets together at a very early age but he soon gave up trying to draw because the results did not meet his high critical standards. He remarks,

"I can't draw but I certainly can see, and I think a lot of my drive to analyze morphology (the forms of things) perhaps comes from being able to see these forms so clearly and yet not being able to satisfy myself by drawing them." Like most of these biologists, he thinks chiefly in pictures, and this seems to go way back in his life.

"The other thing that fascinated me most as a child was mathematics and it occurred to me when I was about 10 that you could have a number system with another base so I invented ones with 11, 12 and 13 bases. I never found out that the 12 base would work better. I worked with the 11 base quite a little and found I could do some problems with it. My teacher caught me and tore it up, but I took that as a matter of course. I had a very poor teacher in high school math. He was the coach and he was the kind who couldn't do original problems in geometry. He took them home and got his wife to do them and then laid the notes on his desk with books around so we couldn't see them. I found this out and when he was out of the room I changed the letters on the diagram around. I was going with a girl whose initials were GRB and I labelled triangles for her initials and of course the entire room was in snickers but it never occurred to the teacher what it was all about."

Most of these men had no difficulty in school, although not all of them showed particular promise there. Some of them found languages difficult and others enjoyed the classics particularly, but for most of them mathematics and science were really easy and they liked them. They were rather more likely to get along with their teachers than not, partly of course because they were bright students, and few of them were smartalecky.

"I went through high school in 3 years at the principal's request. He asked father. Father knew there would be problems when my emotional age was behind. I went to college at 16 and had great trouble because I was really emotionally a child. It was an agricultural college and had lots of boys from small towns and the entrance age was higher than now. I had all this terrific energy and a great interest in people and these things. It has prevented me from having any nostalgia for old college days. I didn't get into any fraternity so I founded one which became a stinking political influence in my alma mater. When I went to college I already knew I wanted to be a botanist, but I'm one in spite of my undergraduate work. It was sound stuff but not very attractively taught.

"My chief problem in college had been social adjustment. In an ordinary family it would not have been so hard but the rest were socially brilliant. Everyone adored father; my mother was the reigning dowager and my brother was always president of his class and all that. But here was I, very calfish, especially seen against the background of my family. I tried to do things the way they did, that was just it. I'm like my father in some ways and like my mother in some so I couldn't do things the way either of them did. It's just experience that has taught me I'm different. I didn't feel different. An old lady I met when I was in the navy taught me to just go ahead and be myself."

"At college I would have rated below 100 in popularity in a class of 150, but in graduate school I was very happy most of the time. My thesis was not particularly interesting; the topic was largely a question of what was available. I don't think it did me any harm and it did give me an insight into a certain field. The main thing was that they left me alone."

This problem of social adjustment is a very common one in this group. It is, of course, likely to arise for any boy, or girl, who is intellectually advanced, goes ahead in school and is thrown with older classmates. The stage of physical development may itself introduce special problems setting him apart from his classmates; even just smaller size or lesser strength can be a handicap. When a boy also has intense personal interests not widely shared he is likely to be very lonely unless he finds another boy or two like himself. Then there is the whole problem of girl friends. Henry had a girl friend in high school, but this is rather unusual in this group, and was rather tentative, anyway. Only one of these biologists was really openly interested in girls at an early age. For example another of them remarks, "I had been a complete misfit in college my first two years. I had been very slow developing physically and this made for difficulties. The third year I got into a fraternity. I did not fit very well there but I had to go to dances and so on. When I came back for my last year, after a year away it was a more social life and I fitted in much better." Some of them had well-concealed crushes during adolescence which they felt unable to express in any way and it was often the case that their first friendship with a girl was with a fellow graduate student. Such problems are not confined to scientists, and are not too unusual. I think some changes in educational practice could be a help here.

Unlike most of this group, Henry had known from a very early age not only that he wanted to be a botanist, but that it was possible to make a living as one, and he knew, too, how to go about it. But he, too, mentions the importance of being left to work on his own.

99

The story of how John became a geneticist is quite a different one in many ways. He grew up on a small farm on the edge of a small town. There was an older brother and a younger sister. His mother had died when he was about 4, and most of what he remembers of growing up is a series of housekeepers. He had a good deal of freedom as a child, no one ever knew what he was doing. Of course he did have some chores, but only a reasonable number of them and he did not mind. He had a lot of friends in the neighborhood and spent some time in Boy Scout activities. He says, though, that he never went around with girls until he was in college. Before that he got crushes on them but they never knew it. He does not remember much about grade school days. There was a high school in town, with about 200 pupils, and he went to that as a matter of course. He had no particular plans, except that he rather supposed he would become a farmer. About his schooling, John says,

"I wasn't phenomenal in school at all. It's a funny thing but the older I grew the more I stood out as a student. I wasn't at the top of the class in high school but my first year in college I was at the top of the college. Of course there was some factor of interest there."

This must largely have been because of increasing interest and application on his part since it is more often the other way around. The top man from one high school will meet top men from other high schools in college and the competition is much tougher. Only one of them can maintain top position, if any do.

"In high school I was interested in physics and chemistry. I suppose the biggest influence in high school was a particular teacher who taught physics and chemistry. I guess she didn't know too much but she was a very good teacher, allowing people to go ahead and express an interest. She used to let us work after hours in the lab and fool around and it's a wonder we didn't blow things up. She thought I should go to college. I wasn't hard to convince although my father was hard to convince. I had to browbeat him. I finally just told him I was going so then it was all right."

It is very unusual for the parents to object to a college course. It was true that a number of the parents did not think of sending their sons to college but most of them, once the issue was raised, were quite willing for the sons to go and helped them financially when they could. A large percentage of all of these men earned at least part of their expenses and a number were entirely on their own.

"One reason it was easier for me to go to college was that I had an older brother, 8 or 9 years older. He was a very smart guy, but he died at 18. Father wouldn't let him go except to a small local college and I think that's why he didn't argue with me as much. I worked part of my way through, not all, because I had a small amount of money left from my mother.

"I went to an agricultural college because I thought I was going to come back and be a farmer. I liked English a lot, but as a sophomore I got a job reading papers in agronomy. Shortly after this the top man in the group came back from getting his degree in genetics. He was all steamed up about it and gave a course in it. I took the course and found I could do the problems better than he could so I got interested in genetics by taking his course and also working for him at 30 cents an hour classifying wheat.

"The agricultural school had a supply of all kinds of grains for high school collections. This was a business and I was put in charge and enjoyed it. I identified, collected, ordered, and all the rest of it. It was a kind of established business and passed along. In connection with that every summer we grew a little museum plot.

"When I got interested in genetics I did some crosses, just for fun. I had kept bees and rabbits on the farm but had no particular interest in them from a genetic standpoint. But as soon as I learned about genetics in college I was very interested and I used to read about it in my lunch hour.

"I was an agricultural major from the beginning. I suppose my major subject was agronomy but another dominant interest was insects. I spent lots of time making drawings of insects. I took all the courses offered and the professor suggested I should do a special problem, the ants of the state, and this depressed me so I quit being an entomologist."

Why this should depress him is not clear. Perhaps because the "professor was a very scholarly fellow who sat in an office all day."

"Then this professor I had been helping said he would get me an assistantship at his college for graduate work and he did after I took my M.A. with him. My teaching assistantship was in agronomy and I was supposed to study grasses. But I soon found out it was some one else's problem and I was just doing the work, so I decided I'd be a geneticist. I had had a minor in genetics from the beginning, and I probably changed because of the personality of that professor. He was a very charming man. I didn't have an assistantship at first after I changed but a few months after that the genetics professor gave me a research job. I liked it because he didn't want to be bothered and he was the kind of a fellow who was very generous scientifically. Instead of saying 'Here is my problem, do so and so on it,'—of course some times he had to,—

but in general he'd say, 'Here is a problem, work on this' and then it would be your problem, not his. It was a very generous attitude. This has had a big influence and this is the tradition in genetics."

John points out that much of his career has been determined by accident, not by any planning on his part. The effect of personal contacts, and of the spontaneous help of interested teachers is particularly clear in John's case. Even so, it is not accident, only, that it was John they helped, nor that as soon as he even heard about genetics, "I became so interested that I used to read about it during lunch hours." With no idea, to start with, of the possibility of a career as a geneticist, he followed his immediate interests, and a series of easy steps took him from scientific farming to research in genetics.

The story of Ed is quite a different one. As you will see from his own brief account below, his interest in science apparently came suddenly, from one experience. It has been great enough to carry him to a pre-eminent position.

They lived in a country town where his parents had a newspaper. He remembers that there were always many books in the house, but he can give the name of only one, called *This Wonder World*. His parents had never put any pressure on him for grades, nor did they have any special vocational plans for him. Most of his early days he spent in out-door activities, and he went to college largely because of his interest in athletics.

"I had no interest in any scientific things as a kid. I did like high-school chemistry and I liked chemistry in college, and majored in it and math because I enjoyed doing them and did them well and easily. Latin was very difficult and modern languages difficult. I could get them if I studied, but it involved proportionately much more work. My chief interest was in athletics and football, so much so that summer work was of a nature which would develop me physically.

"The transfer came in my senior year at college when I was taken over to the state university, principally because the chemistry professor was interested in locating a teaching assistantship for another student. I had my eyes opened then for I saw students doing real research work and I had an opportunity of talking with a very outstanding professor of chemistry. Just an afternoon visit. It was the research that interested me, the idea of using chemistry to find out new things. So I applied right then for a teaching assistantship which I didn't get, but I went to summer school there thinking I'd like to get going and find out what it was all about. I must have been very excited. I had worked hard, I suppose, for that fall I was given one of the quarter-time teaching assistantships. My family was not wealthy and that $30 a month meant a lot. Of course what I wanted was the $60 teaching assistantship, which I got in about six weeks. As soon as I could I got started in research activities. Some were started the first year and during that year I worked off many of the required courses so about half my time the second year and all the third I could spend in research."

I wonder, if Ed had not just happened to make that trip with his professor, if he would have continued in athletics? One has the feeling this would have never have satisfied him if only because he clearly was capable of so much greater a role in the world. He could not possibly have seized upon this experience, though, if it had not met something he was wanting without knowing it. He seems to have known at once "That's for me," and he was right.

IX

Becoming a Physicist

HERE ARE the stories of several of the men who became physicists. Since the theorists and experimentalists are quite unlike in some ways, I shall include both. Again it is true that some of them knew quite early that the physical sciences were a vocational possibility, and others did not hear of them in such a connection until well along in school. You can know that there is a school subject called physics, and men who teach it, and you probably will have learned that there have been famous men called physicists, who found out certain things about the world, but this is very different from realizing that you can make a living at finding out things in this field.

Martin was the son of a consulting engineer, who had had some college training. His mother had worked as a reporter for a while after she finished high school. He says,

"I can't remember much about grade school except the fact that I got reasonably decent grades right along and that I was fairly interested in science and mathematics. I had a friend in 7th or 8th grade who was the son of a druggist and we got a chemistry set between us and played around with it and almost blew up the house. We spent our spare time memorizing the table of elements. I never got along in languages, I couldn't see any sense in memorizing grammar. In history I read so much I had many more facts than the rest whether they were right or not. I think probably the interest in science was partly because of father. When he was home he liked to do shop work and I used to do some with him. He was rather meticulous and in some ways this was discouraging for a beginner."

Several things about this statement are very characteristic of theoretical physical scientists. All of them liked school. Most of them preferred mathematics and science to other subjects. A number of them spoke of dabbling in chemistry, and of still being surprised that they had not blown up the house, and many of them did other sorts of things with their hands, such as the shop work mentioned by Martin. His mention of memorizing the table of elements reminds me of another of this group who became interested in mineralogy when he was a boy and who papered his room with sheets of paper on which he had copied tables and descriptions of minerals.

Martin goes on to say,

"I was rather sickly. I imagine it was more allergic than anything else, although it was not recognized at the time, and I was out sick two or three months each year. One term in high school I was only there for a month. It was always something special; my brothers and sisters always had measles and things like that but those never bothered me. I had tonsils and adenoids, hay fever, a mastoid, and appendicitis. This meant that during most of the winter months I didn't get out and I got to reading fairly early. Since I was in the 8th grade I've been in the habit of reading 4 books or more a week. I read pretty much anything. If I'm working hard in physics

I like to relax by reading history or almost anything but physics. One spell in high school, when I was sick for three months,
I decided I was going to go into history and I spent the time
in drawing up a historical chart beginning with the Egyp-
tians."

His frequent illnesses, and his omnivorous reading are also
characteristic of this group. There were only three who had
had no serious physical problems during childhood, and all
of them read intensely and almost anything they could get
their hands on. Two of them remarked that they thought they
got their first interest in science from reading science fiction.
Reading, of course, is not a very social occupation, and the
physicists, like the biologists, rather tended to be quite shy.
Martin, however, is unlike the others in that he got over this
rather suddenly, although not very early.

"I did very little going out in high school. Mother was very
worried about it. I felt very shy. I started in my junior year
in college and all of a sudden found it interesting and easy
and rather overdid it for a while. Let's see if I can remember
how it happened. I just happened to get in with a group of
fellows and girls who were interested in artistic things. I
started going to the symphony concerts at that time and we
got in the habit of going Saturdays to Little Italy and sitting
around and drinking wine and talking. Since that time it's
been a thing I could turn on or off at will. There were a number of periods before my marriage that I did a lot of running
around and other times I'd be too interested in something
else. I've always been self-conscious at social functions and
never cared very much for them. With a few people it's different."

In high school one of the teachers had great influence on

him, and this experience oriented him towards science at the same time that out of school experiences convinced him that he did not want to be a business man. Not all of these men had occasion to spend any time in commercial activities, but quite a few of them did, usually in the course of making enough money to go to school. None of them liked business except one of the biologists who found it of interest but was glad to go back to science. The extreme competitiveness, the indifference to fact, the difficulty of doing things personally, all were distasteful to them.

"The first few years in high school I don't remember anything special about, except that I managed to get fairly decent grades in mathematics. I took physics and didn't like it. I had taken chemistry before I got there, but there was an extra course that sounded interesting so I took it and it turned out there were only four students in the course and a very interesting teacher. He sort of took personal charge and let us do pretty much what we wanted except that he was extremely insistent that we take care and do a good job. We worked through all of analytical chemistry there and I got a feeling for looking for small traces of elements, etc. This convinced me that I wanted to be a chemist. A little earlier I had gotten a job with the phone company which was with a fellow studying to be a chemist. I read Slosson, *Creative Chemistry*. This was the romantic thing to be. I think that teacher had more individual influence on me than any other."

Some firm, apparently interested in increasing the supply of chemists, had sent *Creative Chemistry* around to a number of high schools, and it seems to have been a very successful promotion. At least several others of my subjects mentioned having been influenced by it.

"When I was still in high school I took a job one summer at a Yacht Club. It was a navy camp and one of the instructors had been a radio operator. He got me interested in radio and we played around a certain amount. That winter he and two other radio amateurs decided to open a small radio equipment store in town and they asked me to go in. Perhaps they thought father might help. Dad did put up some money and we opened a small store and for a while I spent part time there. When the craze hit in 1922 or 1923 the place was about swamped, it was the only store in town. What was made on the store pretty much paid my way through college. While this episode was interesting I was pretty sure I didn't want to go into business. You always got essentially people fighting you. During part of this time in addition to working at the store I had been a part-time radio writer for one of the papers. While that was interesting, too, it didn't appeal as a life work either. By then I was convinced I wanted to go on in academic work.

"College was actually pretty much taken for granted. My mother was convinced from the beginning that all her children were going to college. I just went to college expecting to be a chemist. I had no very special idea about it. Two things happened in my freshman year. I took the college chemistry course plus the lab course. The lab course threw me for a complete loss. I think it was taught by a poor teacher who was careless of the reagents and they weren't pure. I got traces of everything and reported it. I didn't like the way the course was taught because I was told everything I was supposed to do and it soured me on chemistry.

"I got acquainted with a young man who had just come there as an astronomer and was teaching mathematics. He was perhaps the most inspiring teacher I had. He let you go if you wanted to go. I needed some money so I helped arrange the library and so I had a chance to look over the mathematics books. At the end of the year I decided the devil with chemistry, I'm going into physics.

"At that time the college had a course in physics which was not popular. My class had three students and this gave us personal attention. I thought of going on with it. My father was very dubious about it. He wasn't sure that physics was a thing you could get along with but he didn't push it very hard. He talked to me about it once and said, 'You will have to go on in university work and won't make any money.' I said I knew that and he said 'If you realize it, that's all right.' There was nothing special about the course except at the end of that year a prize examination was given. At that time physics was taught practically everywhere without the use of calculus and still is in many places. We didn't get calculus until our sophomore year in mathematics and I still can remember the annoyance and the feeling of being cheated out of an extra year or so of activity by not having had it earlier. At any rate the physics course was given with the calculus but didn't use it. So about the middle of the second term I got disgusted and decided I wanted to learn physics the right way and asked the teacher for a text. He smiled and gave me one and I studied that so when the exam came along I gave it all in calculus and got the prize. This confirmed me, of course, and the next two years were extremely pleasant. I divided my time pretty much between astronomy and physics. There were just three of us and we'd go to the professor and say we had finished up this and what should we do next and he would say, 'What do you want to do?' So we'd tell him and he would give us manuals and get the old apparatus out and usually it would have to be cleaned and fixed up, and he would tell us to work it up and we would have a fine time.

"My teacher felt I should go on to do graduate work. This was kind of a surprise to the family and a little bit of a worry because my brothers and sisters were coming along and there wasn't too much money. But I applied for scholarships at three places and took the second offer. My main danger the first year was to keep from galloping off in 24 different directions at once. I found it extremely interesting and exciting. I started work on an experimental problem, but then I would get an idea for a theoretical paper and work on that for a while, and then go back to the other.

"I think my teacher in high school had given me a few nudges in the direction of research. Both the professors at college with whom I was in close personal contact and saw daily were active in research themselves and I just soaked that stuff up. I find it hard to think back to the time when the idea of research and just spending all the time I had available on trying to understand anything wasn't just there."

The story of George, who became an experimental physicist is quite a different one, but it is fairly characteristic of the experimentalists. He did some manual things as farm boys do, but was not particularly interested, and he did not have radio sets and gadgets of one sort or another. Farm boys didn't then. Nor did he do any particular amount of reading. So far as he knows none of his family had gone to college before him, although some have gone since; his father had had about a 6th grade education and his mother one year of high school. He started out in the usual 7 months country school, near home, but his going on was unusual. He says,

"My father and mother were rather an exception in the community which can be pointed out in this way. We lived out in the country about 7 or 8 miles from a high school. The country school to which we went was very close but when I finished seventh grade the school was having its usual ups and downs and the high school was no good. So my father and mother decided to send me to another school and it required boarding me away from home, and that was quite the talk of the area, that they would waste money boarding me.

"My recreations were the usual ones, physical activities. Whereas most parents in that neighborhood believed that children when not in school should work along with the hired help, both father and mother adopted the attitude that they expected me to do a certain amount of work but didn't care when I did it. They would lay out a certain amount per day and if I wanted to get up and work hard and be through with it that was up to me. That was always criticized because I was always enticing the other boys away when they were supposed to be at work. I earned the title of being one of the laziest boys. Father required only that I do my work and do it well. He did this with the other help as far as possible, too, like piece work. From that I learned how to make time on manual things and at the same time to do as well as required. But we had no tools and I did no carpentering. Up until I went to graduate school I never knew I had any ability in that respect at all. I didn't do a great deal of reading. In those days the books that were available were novels and I wasn't particularly interested.

"I think I wanted to go to high school. At least I was perfectly willing to go. It came rather suddenly. I don't think very much was said about it until possibly a few days before I went. I suspect my mother had more to do with it, she had thought it out very well, but I don't think she said much even to father. His reaction was that as long as I did well he'd help me go to school. If I failed I could come home and work. He always thought farming too hard for anyone and that anyone who had intelligence would get off it. The first year or so was pretty rugged. It was difficult to find a satisfactory place to stay. We had one little course in physics in high school, not a lab course, and the usual mathematics. I think I was probably the top of the class in that.

"There was an incident there that has always been amusing to me. The only time I had any trouble in school was with the physics teacher. About the middle of the year she was showing how the water level in the boilers was determined. She left the gauge open and I said all the water would go out. The argument got hotter and hotter and finally I volunteered to show her, at which time I got thrown out of class. There again it was what father always said, you have to think things out for yourself."

This is the sort of incident that can happen when a teacher (or parent) is so insecure as to be unable to tolerate the suggestion that she might be mistaken, or might lack some particular piece of knowledge.

The experimentalists are like the theorists in their early preference for mathematics and science classes, and their disinterest in languages, and difficulty with them is somewhat greater than that shown by the theorists. Very few of the experimentalists were avid readers. The teachers at George's school were all college graduates, and the principal talked a good deal about going on to college. George was early determined to go. He liked school work, he did not like farming, and he had some idea of going into medicine. He tells how he happened to think of this.

"I started out for medicine. Along about the time I was 14, there was a young doctor came to the community and he boarded in my home. I used to drive a car for him and I got rather interested. My real interest got started from an incident one afternoon when a colored child had gotten badly burnt. Neither parent could hold the child and a neighbor couldn't do it either so he came out to the car and asked me if I thought I could hold the child and give it ether. It was badly burned. Apparently I succeeded because that night he told my mother she had a young surgeon in the family. Maybe that started it, but when I went to college I intended to go into medicine.

"I went to the nearest college. The medicine idea shifted gradually. Two things happened, I think, that caused a shift. One was that by pure accident, in the first year mathematics course I was lucky to be in the section of an exceedingly good teacher. I always liked to be in the back of the room if I could. It seemed that during the first week this professor would start asking questions and begin at the front end, and by the time it came back to me I would have been able to get the answer, from the book or by working it out. Then he began another trick, if he didn't get the answer on the first three or four he would say, 'How about my old standby?' and call on me so I felt I had to know it. From that he began to take quite an interest in my work and before the year was out began talking about my working up the second year for myself during the summer. So I promised I'd try and he said he'd give me an examination in the fall and then I could go into the third year which he taught. I never have known if I passed it or if he let me by, but I went on with him. He wanted me to specialize in mathematics, and along with that there happened another incident.

"I had become engaged to my wife and she wasn't keen about being a doctor's wife and undoubtedly that had an influence on me. She wanted her husband at home a reasonable amount of the time. As it turned out, especially during the war, that isn't just what she got. So I gradually drifted in the direction of mathematics. The second summer I worked up some other courses and at the end of the third year had completed four years of mathematics. Along with it I took one course in physics but I wasn't particularly interested, and I had one year of chemistry. The last year I found all I lacked for a B.A. instead of a B.S. which wasn't considered as good a degree, was a year of Greek so I took that. It was a kind of training that to my mind is lacking today. I even wound up with the highest grade in the class.

"The idea of going on to graduate school came from this math professor. When I started I only intended to go through for an M.A. I didn't see my way clear further. This professor helped me to get a fellowship and that plus my father plus my wife's working made it possible for me to go. I started out intending to spend a year and a half and get an M.A. and go out teaching in mathematics.

"Then again one of these things happened. The first summer I took two courses in mathematics and for some strange reason I was assigned a course in physics. The two courses in mathematics were taught by two foreigners and they were the two most discouraging courses I've ever had in my life. One in particular was taught by a famous English mathematician and he was teaching completely over our heads. I thought it was my own dumbness. I worked as hard as I ever worked in my life and accomplished as little. A few days before the exam I mentioned it to one of the other students and he was feeling the same way. So the next class he had the nerve to go in before the teacher came in and he went up front and asked and pretty soon he discovered most of us were in the same boat so when the professor came in we stopped him and told him this. He asked around the class and they mostly said the same. He had assumed we had had two years of mathematics that we hadn't had and so he gave an exam I could have passed in high school. I was thoroughly disgusted with mathematics. The only course that was half decent was the physics course but I wasn't prepared for that.

"At the end of the summer I thought I wouldn't go on with graduate school and I decided to go down town and get a job. If I still felt the same way I'd just continue working instead of going back next term. I got a job as a salesman. That was another lucky stroke. I went down and started putting the same effort into that. I began selling boys' shirts and I'd never bought a shirt in my life, mother always did. So I went to the library and got out three books on cloth. I read two that night and by the second day I understood a little more. I thought that if you wanted to be helpful in selling and it would be your job to learn what you were selling and it paid off as far as sales were concerned. Of course then it was said I was a sales grabber so I was told to take my turn. I said that was all right and did take my turn but I still maintained the highest sales, but it was because by then I was selecting out the good quality. I got called down for that, and they said there would be a lot of returns, but I asked them to check it and there were hardly any. Then I had a run-in with the buyer and was transferred upstairs to sports goods and the same thing happened there. It was the same old trouble. No one ever bothered to study their stuff. At the end of the month I saw very clearly that in an industrial job you didn't get anywhere by knowing more or doing more than anyone else. By that time I was convinced that that side of the world was a pretty sorry one.

"By then I had also decided I didn't want to go on in mathematics. That one course convinced me that physics was what I wanted. I had my fellowship transferred and had a long fuss with the Dean who wanted to assign courses and I wanted to work up to them. So I started out from there and with essentially undergraduate courses.

"I liked it very much better and I found I somehow had time on my hands and very soon I wanted to try my hand in the lab. I had never had any tools in my hand. Again I had a lucky break. I went down and told the professor and said I'd like to try and I'd be glad to begin by opening boxes or anything else. He laughed and said as it happened there were a lot of boxes to open and so he put me to work. Presumably lying dormant in my fingers was an ability I didn't know I had. Within a month I challenged him that I could make an electroscope work better than he and I won. I've always wondered if he let me do it; he never would admit it but I would not expect him to.

"I found that almost anything in experimental work I had no difficulty in doing. Glass-blowing and so on just came to me overnight. I learned mainly just by doing it. Machine work was all pretty much the same way. Handling the tools just came naturally as if I had been doing it for years. So much so that when I came here and took over the shop I said I'd never ask them to do anything I couldn't do myself. At first they sometimes said they couldn't do things, but I always showed them and since then there hasn't been any question."

It is rare to find any planning ahead in the early years. Mostly the men just go from one thing to another, as occasion offers. The next story is particularly interesting from this point of view. He had an early bent to mechanical things. He went to college, largely because of his mother's dreams for him, but even there and after he had courses in physics, it was some time before he found out about research. His story is a particularly good illustration, too, of a sort of unconsciousness about many aspects of living that is not uncommon at the college years, and not unheard of beyond them. Ernest described himself to me as an experimentalist but one of his colleagues once told me that his greatest contributions had been theoretical.

"I really can't say when I got interested in things mechanical but it's just about as early as I can remember. About 6 or so I was interested in pretty much anything electrical, the usual things that kids are interested in, autos and so on.

"Father never got even through high school and started at practically hard labor at 13 and got from that to be a star salesman. I don't know when he found time for the things he did. He was quite athletic and at that time there were amateur athletic groups and he was stroke. I never realized how good he was at the time but later I found some old papers and found that his crew was the best anywhere around. All the training was done after a day's work. Then some time later some of the books I read when I was a kid were some Inter-

national Correspondence School texts on engineering which
he had studied. That's a lot of work when you are working
hard too. Father was a better man than I was or ever will be.
Even when I was young and strong, my father was much
stronger and tougher than I was always."

References to parents show marked differences in the attitudes of the sons. Ernest's respect for his father was very
great, and this is generally characteristic of the physical scientists. It is less characteristic for them to have any great feeling of closeness to their fathers, or great affection, but Ernest
and his father seem to have been very close.

"Father had a strong mechanical bent and I learned quite
a bit from him without realizing it. From the age of ten or so
I was entrusted with keeping his car serviced. By the time
I was 12 there were several of us interested in radio and we
made a set. I was sort of leader and I did most of the design-
ing and construction, the others did the operating. This was
a transmitting and receiving station. I was always sure I
wanted to be something of the engineering sort. I had never
heard the word physicist, of course, and neither had either
of my parents. I had fairly large sets, Meccano and Erector,
at a rather early age. You can get a lot of action for a reasonable amount of money. The folks would buy motors for toys
and when I got to be old enough to be a radio amateur I was
more organized and then it was mainly a question of making
up my mind what I needed. We had all kinds of complicated
arrangements. For a while we formed a small company to
manufacture transformers. It was sort of a joke. The power
company was putting in a lot of new transformers, and so we
got any amount of stuff given us by the uncle of one of the
boys and then we cooked up a deal with another's uncle to

dig a cellar for $20 or $30 worth of wire, and we made some transformers and sold them. I never worked so hard in my life. We sure found things out the hard way. We had considerable instruction but it was practically all of it from books and we found out how to do it the wrong way first always. It just happened there were no radio amateurs around who knew more than we did so they learned from us.

"Father never helped me make anything. On the other hand if I asked him how to do something he always knew and he had tools around which he got for his own purposes and which I appropriated so it's hard to describe. He never gave me any formal instruction but I learned a lot. Not about electricity but about mechanical things he was very, very good.

"In high school I took chemistry and physics, all there was of both, about a year of each, and then some odds and ends of surveying and such courses. I took all there was of math and some that didn't exist, i.e. the math teachers were very interested in me and awfully kind to me and gave me instruction in things that weren't really on the books and I learned some on the side myself.

"I got through high school quite young and my folks didn't think I ought to go to college quite so soon so they sent me for a year to the technical high school there, so I had perhaps better training than ordinary in that way. That was a well-run course. I spent most of my time in the machine shop.

"Going to college wasn't taken for granted. My father was the son of immigrant parents and had his first job as a blacksmith, so college tradition in the family wasn't strong. It was mother's idea. Her father was a minister and she was of a fairly well educated family. Among my boy friends none went to college. I always had had a good time in school and would just read anything. I wouldn't say I liked all my studies but I liked anything scientific or mathematical and was all in favor of more school. Father was all for it but it was mother's idea in the first place.

"I got a scholarship and went to college intending to become an electrical engineer that being the nearest thing we knew of to what I was interested in. Then my money ran out and I went home and continued in the college there. About then I had to take sophomore courses in physics and the professor thought well of me and he said, 'Why don't you go into physics?' It seemed a lot of fun and he thought he could stir me up a job at another college and said there wasn't much difference between the physics and the electrical engineering courses and I could change back if I wanted to. I guess he must have done some considerable wrangling but he got me a job as assistant when I was a junior, and I came up here and thought that was a lot of fun.

"I was pretty young and I guess not any too noticing about some things. I didn't realize there was such a thing as research either at that time. One fine day I was downstairs and saw someone wandering down the hall with a soldering iron, something I recognized. He was a graduate student and didn't look like he knew what he was going to do so I went with him to help and spent most of my junior year working on his research and had a high old time working on it.

"This was a small place in those days. No one told me how things ran. I didn't know about any of the places where people gathered. I'd seen this fellow around the teaching labs but I'd never heard of the idea of research. I'd taken courses and I thought that teaching was what professors did. The fellow I assisted for was one of the few that did not do research and I just saw him in his teaching laboratory. I didn't have any idea of what the student I helped was trying to do. I could see he was building things that he didn't know how to do and I did so I helped him for the fun of it.

"There was an International Research Fellow here. He's a smart guy but pretty excitable and not dependable. By the time I got to be a senior it got to be recognized that I was pretty useful in the lab so they gave me to him for research associate and by that time it got time for me to graduate and I began to wonder what to do. This research Fellow was offered a job elsewhere and he could bring along anyone he wanted so he asked me if I wouldn't like to go and I said sure. The next day I ran into the department head and told him this and he didn't say anything about it, but after a couple of weeks passed I got an offer of an instructorship here and that surprised me and I accepted. So I stayed here to get a Ph.D. I was only 20 and just had hardly grown up yet. I took chemistry too and got along well in it and had a good time. I'm sure I would have been happy as a chemist only I just had more experience of thinking mechanically that made me seem to fit into physics better.

"As it happened I worked on several problems at once, but the one I did my thesis on was a joint paper with the head, so he really suggested the problem and I just worked with him. It's a very rare student that can tell a good problem when he sees one, can start it off and carry it through. I certainly couldn't have."

X

Becoming a Social Scientist

HERE ARE the stories of two of the psychologists, and of one of the anthropologists. I have chosen them to illustrate the different early interests most characteristic of psychologists and anthropologists. Children are even less likely to know of psychology and anthropology as possible careers than they are to know of biology or physics. One of these psychologists expected for a long time to become a writer. This is characteristic of many of them. The other whose story appears here intended first to become a minister, and much of his motivation throughout has been at least in part a service one. This is not uncommon among psychologists, particularly among clinicians. Neither of these situations is likely to arise among men who later become biological or physical scientists, although both may appear in the histories of anthropologists. The anthropologist whose story is told here, however, was first interested in natural history. This is true of a number of anthropologists but unusual among psychologists.

Stanton began talking about his family background, which is similar to many in this group.

"My family background is both farm and a fairly good upper middle class group. They had farm, mercantile and professional affiliations. In the professions they were doctors and lawyers. My father is a very able, hard-working man who put himself through college after he married, went on to graduate work and ultimately became a professor. I am surprised to find myself teaching now. In adolescence I hated the idea of teaching and hated educators.

"We moved around quite a lot. I never was particularly interested in school. I was one of the good boys and I got along nicely in school because I was good. I was one of the rather non-social shy children whom teachers all like. My scholastic interests were minimum, except English and drama.

"I always had one or two close friends in early school days although I didn't get along too well with other children. Then we moved and I had a difficult year. I was a stranger, and I was very tall and thin and physically ineffective. I found out what it meant to be a minority member. And then we moved back and sex had arrived, which was very happy. My high school years were very highly heterosexually oriented with lots of dating and dances and great interest in reading and writing. I did a lot of acting and journalism. I played the piano and learned to play jazz. I was thoroughly and completely an ingroup member then."

This happy interest in dating and dancing is true of many of the social scientists though not of all of them. More of them were interested in music than of any of the other groups except the theoretical physicists. In the number of them who had physical problems during adolescence, they are also rather like the theoretical physicists.

"It never occurred to me that there was anything anyone did except go to college. I don't think vocational plans ever entered my head. The things I was excited by were creative writing and courses here and there, one in Herodotus and one in archeology. But by my junior year I had decided to be a short story writer. In my senior year I was president of the English group and associate editor of the college magazine. I became quite a Bohemian, interested in the esoteric and an expert on metropolitan speakeasies, and on local wines. I met a short story writer of some stature at a summer resort and he invited some of us to his room and would read us his stories. I was terribly shocked and very much excited, and thrilled, and quite happy to think I was sitting talking about such things and being very adult."

Quite a surprising number of the social scientists were active in college literary societies and glee clubs and president of this and that, and this was true of very few of the other groups.

"At the end of my junior year I got engaged. She'd had to go to summer school because she'd been out so I did, too, and we took Psychology 1. We sat in the back row and held hands and went out and studied together under the trees. This undoubtedly had a profound effect on my interest in psychology. On the other hand we also took sociology together and it didn't have any effect on that. I looked on psychology as a refuge from the vagueness and what even then struck me as the amateurish guesswork of English criticism. I was quite dissatisfied with the balderdash about motivation and character that you got in that.

"And at that time I had begun to suspect that I might not be the kind of short story writer that eats. I kind of liked learning theory and I loved the objective questions. Just this incredible number of isolated little facts. It must have hit me as providing a certain solidity. It was so nice to get things down to a really precise point. That's something you will see in my research. I take perfectly good problems with lots of phases to them but I always end up measuring. I majored in psychology after I fell in love, after I had gotten bored with English and the people in the English department, the kind of aesthetic pseudo-intellectual they had there, and I found some very smart, sharp, good people in psychology who were very much interested in me and who spent time with me and gave me interesting things to do, like measuring things and finding relationships. If any one course were to be given credit for my final choice, it would be the course in experimental."

This shift from English and literary criticism to psychology, and the description of what he liked about psychology is almost as apt for a number of the others. They had tried to find out about people, through reading and through writing about them, but this became unsatisfying. At the same time they learned about psychology which gave them a more direct attack upon the problems that bothered them, and one which also satisfied another desire most of them had, for regularity, for measuring, for doing something specific.

"I decided to go on for graduate work. I went with the notion that I would go only for a year and it would be nice to get so far away from home and fun to see something of the rest of the world. I suppose this was probably about 80% of a decision toward a profession but I can remember I had a feeling of not having made a decision. But I couldn't seem to think of any alternative. I didn't like it; it was ghastly but after two months you couldn't have gotten me out of it. I guess it was like taking religious orders. Part of it was the enthusiasm of some of the people. I started working with one of the professors and started working on an experiment. I did a lot of research and then there was just no further question.

"During my senior year in college I was very highly motivated. Just where that interest came from, whether it was derived on a semi-conscious level, that I was working on my own problems, or whether it was the subject matter or whatever it was, I was very much motivated. In graduate school I really became, I think for the first time, a thoroughgoing scientist in the sense of being interested in the purely scientific aspects. I was fascinated by conditioned response and I remember the great day when two of us set up an apparatus and got a conditioned response for the first time. I felt as if I had found myself. This was my home. I did a lot of research; there was an attitude more or less pervasive that students should get in and do research. It didn't matter much what they did but they should be doing something. Once in, my professional life has all been sort of automatic. You go on and do what is expected of you and you do it as well as you can. The whole business of psychological science seemed to me glamorous and exciting. It still does. I liked the people I met and it looked useful and something people would approve of. It just seemed a little strange that people would pay me to do it."

Bob's background is quite a different one, in many ways. His family kept itself pretty much apart, and the feeling of family difference from others was very strong. His social life was practically nil until he got to college and there was very much colored by religious interests. In spite of the fact that the family interests were distinctly non-intellectual he was fascinated by books and reading from childhood. One wonders here, if this might have been an escape pattern, and an early and subtle form of rebellion against the family.

"Father was a construction engineer and finally became very prosperous. He had a college education but he was most certainly not an intellectual. He had a good mind but tended on the whole to scorn professors and such and did relatively little serious reading. Mother went to college but I can't remember if she finished. Any expression of her intellectual side has always been very much restricted by her strong primitive religious background.

"I learned to read long before I went to school. I have really no idea how I learned. It's characteristic of the family situation that the first book I read was a fat Bible story book. Reading was pretty nearly all my life in the early school days and writing. I liked all types of school work. Looking back on it I liked them too much. I was a shy youngster. I can't believe I had any social adjustment in the group. I had enough companions at home to make it unnecessary. We always came home right after school. The notion of not coming directly home would never have occurred to any of us. I have often thought my parents managed to achieve censorship in so many fields without its being an obvious thing. You did things at home, you didn't go out. Dancing, card-playing and movies were verboten, and even carbonated beverages weren't quite right. It's kind of an incredible family pattern.

"In our home reading was only sort of all right; if you had done your chores and there was nothing else you must do it was okeh to read. It took me a long time when I got into university work to feel fairly comfortable reading early in the morning. Reading early in the day meant you were sneaking it in somehow. I think most of what I read was pretty sappy stuff but with a fairly wide age range level. It was pioneers, Indians, juvenile fiction, stuff of that sort.

"I was practically always the top of my grade in most everything, but my social life was unbelievably nil. There was never any question about going to college. It was a definite part of the picture for all of us and I really don't know why. Certainly not because of the values of a liberal education, but I think more on the grounds that you prepare a person to get ahead. We went to the state university where father had gone and started living in the YMCA dormitory. Father had been active in student Y and had kept up his contacts and that was part of the pattern of control. It was sufficiently subtle and not protective.

"I knew what I wanted to do. I'd always been interested in nature. It was about the time of Gene Stratton Porter, and her books about moths of the Limberlost figure very much. I started reading about moths and finding caterpillars and feeding them and I got to be quite an authority about night flying things. Then father wanted the farm run scientifically and he had some of the agricultural people come down from the state university. I don't know how I first got caught up in that but I did and that began to fascinate me. I must not have been over 14 or 15 years old when I read Morrison's *Feeds and Feeding*, a great big technical book. I realize now that what intrigued me was the scientific part. There is a great deal of that, agricultural work is good experimental work and I took that in. We fed the chickens so much this month and kept graphs; we would plant a plot of this and a plot of that and figure how much grain we got from each and that it would be so much an acre. I realize now it was not farming that attracted me but the scientific side of it. But I registered in agriculture and had every idea that what I was going to be was a farmer."

It is striking how many times, in this as well as in the other groups, the first intimation of what constitutes scientific work comes through agriculture.

"With this home background I was a pretty religious little prig at that time, active in the Y and so forth. They put on religious conferences and being of a somewhat emotional and sensitive sort those hit me pretty deep, and I don't know whether gradually or suddenly, I began thinking of the ministry. I think it crystallized at a Student Volunteer conference. Some vestige of good sense kept me from signing up as a student volunteer but that conference did decide me that this was the great crusade and to be in religious work would be the thing. I changed my course. I realized a broad background would be a good thing and so I decided to take a major in history. I thought of it as a good preparation for religious work. And then, too, the exploring, the scientific, the scholarly side of history appealed to me more and more.

"Then I attended a World Christian Federation meeting. That was an education! A group of people who by my standards at that time were liberal and broad thinkers went over on the same boat and we had discussions and meetings. That stretched my thinking enormously and pretty well disabused me of a lot of the narrow religious ideas I had. I remember going through quite a bad day or two when I came to the conclusion that maybe Jesus was only a man and not divine. But that determined me to go to some liberal seminary and it strengthened my decision to go into the ministry."

He had some stormy sessions with his family over his liberal ideas, and over his joining a fraternity and learning to dance. They did not disapprove of his becoming a minister but strongly disapproved of his choice of seminary. Family dissension over the son's plans is more common among the social scientists than in the other groups. Fortunately for Bob he had been able to earn a fair amount of money and this made it considerably easier for him. The economic problem at the student age is a very intricate one which merits further discussion later.

"The motivation for the ministry was really a service motivation. It was quite a strong one I should say, and on a fairly abstract sort of level in a sense. I found the seminary an extremely stimulating place. It was a very free atmosphere at that time and they really believed in freedom of thought and inquiry. There was a bright group of students and in no time at all we were teaching ourselves. In courses in religious education we got a good deal of what really would be clinical psychology. That appealed to me very much and sort of shifted my focus to religious education. On the intellectual and philosophical side there was a steady growth in questioning on the part of the whole group of us. I began to take courses in psychology and by the end of my second year I definitely decided on it. I was interested in child guidance work and the service motivation was definitely dominant.

"I transferred to another school which was quite liberal in giving me credit for the work I had done at the seminary. I would hate to record how little I have had that is regarded as essential,—no experimental or abnormal, etc. But I took my degree in education. There was a clinician who had sense enough to get us a chance to work with children in our own

131

way. It was crude but good and gave us a taste of really working with people on a clinical basis. Perhaps I should mention that although I seem to have had this ragged background I did well in my courses and I think I'm correct in saying that on the matriculation exam I was in the top 1% and I think I was the top one.

"Then I got a fellowship in child guidance. It was a fairly tough year. I was still taking courses at the university and at that time they were rigidly objective, emotions didn't count and the Institute where I worked was everything from ultra-Freudian to statistical. That was very fruitful; that was awfully good training. That year I began to realize I have a facility for working with people.

"Then I was faced with the need for finding a real job and took one with a social agency. It gave me a kind of chance that I think not enough people get, that I just got a snoot full of work. I wasn't particularly thinking about what to do next professionally. There was just so much work to do, so many children to see, so many agencies wanting help with children that I just got deeply immersed in the clinical function. I just learned how to work with kids. And there was none of what I feel has so often killed clinical psychologists; there was no one whom we had to be subservient to."

This emphasis on the value of intensive work experience, uncolored by outside domination, is one which should be taken very seriously. It is certainly something that clinicians must have and they should have it early. Lack of it just cannot be made up effectively later on.

This anthropologist, Howard, had quite early experience with Indians, but his own work has not related to those he knew as a child. Nevertheless the background was of importance for him. That sort of personal acquaintance with a more primitive group is naturally not a usual situation.

"Both of my parents were born in the west, of pioneer families. On my mother's side, the family were mostly farmers. My father was a lawyer with a good practice. The men in his family are generally professional and mostly lawyers. Father was raised next to an Indian village. He was attorney for a number of the Indian tribes, and of course I got interested in the Indian background.

"I went to a public grade school, and then to a private high school until my last year when it failed and I returned to public school. After school I went home and played around the house with neighboring children. We played cowboys and Indians and military campaigns. In high school I went into the woods and studied birds. I was very unhappy in my first school days. I was thrown into a big public school in a sort of tough part of town and I wasn't used to that kind of kids. Later I had pals."

More of the anthropologists than of the others had private school experience. This is in part because of the generally greater means of the family, but also because of their social attitudes, although the latter was not as important with Howard as it was with most of his colleagues.

"I did an awful lot of reading. The house was full of books. My father loved books and he had fine books and I read everything in the public library, particularly nature stuff and military stuff, nature books and travel books. I read some dime novels of course, but I don't think I ever went through a period of trash. I used to read with a flashlight under the covers at night. It was supposed to be bad for your eyes to read at night. My favorite sport was to have a good book and a box of seedless raisins, that was wonderful."

Howard's relations with his parents are not typical of this group, as they seem to have been on very good terms throughout. A number of the others had serious problems at home.

"I don't think anyone thought that I would go to college because I was very poor in math and languages in high school and excellent in history and English and things I liked, and I think the family thought it would be a waste of time. At that time I wanted to become a naturalist. I didn't know much about how to become one except that I knew that I would have to go to college. I was particularly interested in ornithology. Father was pleased but he said that he had never heard of anyone making any money as a naturalist but that it was all right and if I became good at it he would help me if he had any money. I kept a diary, lists of birds that I saw, and things of that sort—that was a passion for me.

"I enlisted in the Navy in April of my senior year and they gave us our degrees without taking any exams. I think the Navy experience is important. I think I would have made a good officer but I did not make a good enlisted man because I was not mechanical. The main thing was I was travelling, but we didn't see anything of the world except the ocean. There was an old-fashioned but good library aboard this ship, including Darwin's work, and I read the *Origin of Species* and the *Descent of Man*. That made a great impression on me. It gave me some scientific background for what had been just a collector's instinct and innate love for nature. I got into trouble with the censor writing a detailed description of the birds of the areas where we were then travelling because of course you could tell where we were from what I had to say."

"During the war I first came into New York and the first day I got leave I arrived at the American Museum of Natural History at 8:30 in the morning and I hung around until they opened the doors and I didn't leave until they kicked me out that evening. I don't think I had any lunch but it was just fascinating to me.

"I got along fairly well with the others in the Navy, but I had a shock. I was innocent and at first I was kicked around. They just herded a lot of us on there. I was a good boxer, and while I never had a serious fist-fight, I did box and I was willing to fight and with that I got a place for myself but it was never high because I was not a good mechanic and also I didn't have much of a leadership drive. I think I adjusted fairly well but I didn't make a good sailor. I did what I had to do as well as I could do it, but that's about all. Everyone would have given his soul to get out, it just seemed such a horribly restricted life, the endless, senseless vulgarity and obscenity and never being by yourself. When I first got there I used to write in a diary and I used to read books and everybody thought that was very funny. At first it bothered me the way they carried on about it, but after I took up boxing and made a place for myself they left me alone.

"It was in the Navy I began to see a difference in a very marked way between the officers and the crew, the lines are very sharply drawn. One of the few officers that I got at all acquainted with gave me law books and I began to read those. That's when I decided to go to college. I began to see that it wasn't just chemistry and math but that there were other things that you could get and learn. I saw the difference between what a technical education did and what a college education did."

I wonder if the Navy has changed much between wars. I should think lots of men had had similar experiences in this war, in learning how to get along with the others, and in finding out what an officer's status might mean. But I also wonder how many of them found copies of the *Origin of Species* aboard ship?

"I registered as a pre-legal student and later changed my major to history because those were the courses that interested me the most. I didn't know anything about taking biology. You can't take a course in Darwinology and I didn't know what these other things were. About that time I began to collect birds for the Museum and switched from history to zoology and majored in it. In my junior year I went on a collecting trip. That was fascinating. I was doing professionally what I had done by myself. That was when I became aware that people did it as a business and at least were able to eat some of the time.

"Then a disaster happened to the family finances which made all the difference between just being able to get along and being comfortable generally. That, of course, stopped the funds for my college. I got a job working with an investment firm and what I learned about business!"

Of all sixty-four of the men I studied only one who had had any business experience had found it of interest. He even came out of it with a good deal of respect for business men, but perhaps that was because he was in a managerial position from the start, having suddenly to take over when his father-in-law became ill.

"Father said that if I wanted to go back to college and could work my way he could give me a little money, so I went back. I went to a Museum and asked for a job. It was a very different thing when I wanted some money although I had done lots of collecting for them. The only job I could get was feeding the rats with garbage from the faculty club. At the same time I had taken some courses in anthropology and been interested and I went in to see the head of the department. He didn't know me, I had had a class with him but I was one of a large number of students. I told him about things and asked him for work, and he gave me a job doing some research in archeology. It's amazing. I don't know why he picked me. I wasn't even a major in anthropology at that time but they had very few majors because there was no future in anthropology then. He gave me this work and paid me thirty-five cents an hour for it and that made it possible for me to give up some of the other work. This was a sorting job and as it grew you had to have a sense of style. I grouped the things and he began the writing and he would write some portions and then pass them on to me. To my great surprise I noticed that he said 'we' decide this and 'we' decide that and I asked him who the 'we' was and he said that I was to be an author with him because we were working on it together. Of course that was a thrill and from then on I majored in anthropology.

"Once I really got into anthropology there was nothing more for me. I think that if any motif runs through it's that interest in history. It's history I was interested in in high school and the major aspects of evolution. I had the interest of a naturalist, too, but dissecting left me cold; I was never tempted to go on in zoology. Those three graduate years were without doubt the most exciting of my life. It was a period of life opening up as more than work and sports and books. We were all poor and working like hell. But week ends everyone would get together and we'd have a big party. It was a free and easy and Bohemian life which was a very interesting and stimulating thing."

This opening out of life into so many fascinating ways is not altogether a common experience, but a number of them do speak of it very movingly.

XI

How Scientists Think

How DO scientists think? In this, as in so many things, they
vary. Some of them probably think in much the same way that
you do, and some do not. But what do we mean by the question?

Thinking takes place in very complete privacy, and your
thoughts are your own unless you choose to reveal them.
Sometimes, even when you want to, though, you can not do
so. It is something of a trick to think and at the same time
observe the thinking going on, and then report it to someone else. One difficulty is that you may change the process
just by observing it. Another difficulty is that much goes on
that is extremely hard to describe in words. So the question
of what actually goes on in the mind, and what terms are
being used in the process of thinking seems almost insolu-
ble. We know more, I think, about how some mentally dis-
ordered persons (particularly schizophrenics) think than we
do about how normal people think. People who have worked
on the problem of schizophrenic thinking have tended to assume that they know what normal thinking is, and I suspect
they may have been led astray upon occasion on that point.

139

Over twenty-five years ago, when I was a graduate student, the psychological world was still buzzing with the aftermath of a famous controversy. This was over the existence of what is called imageless thought. One group of psychologists maintained that all thinking took place in images of one sort or another, or combinations of these. For example, the thinker might have pictures or diagrams or some other form of visual representation of the contents of thought in his "mind's eye." Or he might be going through a process more readily described as talking to himself, in which he didn't actually say words aloud, but seemed to hear them said in his "mind's ear." This type of imagery we would call auditory verbal, which is another way of saying hearing words. The other forms of sensation, smell, touch, and so on, are less frequently represented in mental images. It was pretty well established that some people have sharp and clear images in one sense and not very good ones in others; or they may have good images in any number of senses, or they may have fuzzy ones in all. Further, it is not just the sense itself that may vary, but you might, for example, have very good auditory images of words, but not of music. Such individual differences in form and strength of imagery were pretty well agreed upon. The controversy was over the question of whether or not there was any form of thinking that did not require one or another form of imagery, i.e. over the existence of imageless thought. The real reason why the controversy was so difficult, I expect, is that each man tended to believe that everyone must think as he himself did.

Finally it seemed to be pretty well accepted that there is such a thing as imageless thought. This is naturally hard to describe. Perhaps the best way is to say that it is a sort of feeling of relations, unaccompanied by images of any sort. After this psychologists rather lost interest in imagery. One or two attempted to devise some sort of tests that would give a good measure of the amount and strength of imagery, but none was very satisfactory. It is not too difficult, if you have a good subject, to determine what forms of imagery are most prevalent and clearest for him. Galton first started such investigations in the last century by asking people to think of the breakfast table that morning, and note whether they had visual images of the table and food, auditory images of conversation, or dishes rattling, images of the smells of the food, and so on, and which were clearest and came most easily. You can even work out a sort of rating scale of clarity. The trouble comes when you try to apply the same rating scale to someone else. There is just no objective means of comparison. Two men may each say that their visual images are extremely clear and sharp, and manipulable, but how can you estimate the relative sharpness of the two? Even careful description of the images is not adequate for this, as you will find if you try to get several people to describe their mental pictures for you. And how much more difficult it becomes for images of smell or taste!

It certainly did not occur to me when I set up this project to include any investigation of imagery. I doubt if I had even thought of it professionally since I taught a beginning class in psychology. It just obtruded itself into the situation. When I was talking to the first few biologists I saw and asking them how they went about doing their work, I gradually realized that in some ways their minds did not seem to be working the way mine does. They seemed to be able to call up in mind in detail the most intricate visual patterns and shapes, in

141

two or three dimensions, so that they could compare mentally, say, the skull of one reptile (which is an amalgamation of large number of bones, worse than a jigsaw puzzle) with that of another. Some of the things they did could clearly not be done by describing the patterns to yourself in words, any more than you could easily, if at all, describe the cut-outs of a fancy jigsaw puzzle especially when it was all put together. Then, too, they seemed to use visual images in other ways. They thought about intricate abstract problems in terms of complicated diagrams, for example. Finally, one of them, who was working on evolution in plants, described how he could conjure up in his mind an image of a whole forest, with assorted trees and shrubs and plants, and with a clear perception of the underlying geology, and then watch this change as one geological age succeeded another! Practically a home movie.

This really floored me, and I must admit that for some time I was skeptical that I was hearing what I thought I was hearing or whether I should take it literally. But it was obviously meant literally, and I ran into the same sort of thing in subject after subject, although not in all of them. And when I came to think of it I did not see how they could do some of the things they did in any other way.

What really convinced me, though, was asking my husband about it. He promptly said that of course he thought that way, how else would you. This is a good example of how much you can fail to observe about a person you know well. I have known him well for forty-five years, and while I had observed a number of differences in our mental processes (I have extremely poor spatial perception, for example, and his is very good) this particular difference neither of us had noted. I suppose one reason is that we so commonly come out with the same conclusions and attitudes that it had not occurred to us to notice that we had arrived at them by quite different processes.

In such literature as there is on imagery no marked sex differences had been reported and I had no reason to suppose that I was unique among psychologists. Hence it seemed entirely possible that I had hit upon a difference that might be significant to the study. My own type of thinking is predominantly of the sort of talking to yourself. I can conjure up some visual images, but I do not rely upon them in any way when I am working on a problem, and they are not very clear at best.

Of course everyone can and does use more than one form of thinking. This may vary with the type of problem, or with other things. But most people use one more than another and often to the near exclusion of others. What I wanted to find out from each subject was what forms were available to him and when he used them. So I asked them.

This is an extremely difficult question to answer. Try now to think of how you think and you will appreciate the problem. I usually had to ask leading questions in order to get across to my subjects just what I was driving at, and I often had to describe various forms so they could judge which their own thinking most resembled. The data I accumulated are not very satisfactory because of all these gimmicks, but in spite of that they make quite a lot of sense and I think they are important. I hope they will stimulate some bright young psychologist to devise more objective means for classifying people in this regard.

I got along fairly well with this inquiry with the biologists, and was so concentrated on the differences between predominantly visual and auditory-verbal thinking that I did not pay enough attention to their use of imageless thought. At least I think that is the most likely explanation for the fact that I did not record its presence as often among them. It is harder to perceive in yourself and to describe, but I am convinced that practically everyone uses it to a greater or lesser extent. Some of them gave such descriptions as this: "I just seem to vegetate; something is going on, I don't know what it is"; "I often know intuitively what the answer is, and then I have to work it out to show it"; "You feel it in your guts."

When I got to the physicists, though, one of the first ones I interviewed had me on the ropes. I simply could not understand what was going on in his mind from his description of it. It finally occurred to me, however, that he was talking about imageless thinking, as he was using it for physical problems, and that it was this content that kept me for some time from seeing that the process was one familiar to me with other content. It is precisely the same process that I use when I am thinking "clinically," when I am getting the feel of a person and how he works. This is not a verbal process, even with me, although it may get translated into verbal terms when it is finished.

This raises another point. Since we communicate with one another chiefly with words, most thinking has to be translated into this form before it can be passed on, either in terms of speech or writing. This is often difficult for persons whose mental imagery is not usually in words.

I thought the physicists were difficult to understand when they undertook to describe how they think but the psychologists were worse. Partly because of greater sophistication they were much more meticulous and much less certain in their statements on the whole. Furthermore a number of them kept talking about kinaesthetic elements, that is, feelings of muscular tension. I think these may not be images, but actually barely perceptible sensations which could easily accompany the process of imageless (or any other type of) thinking. A man concentrating deeply may easily get into a strained physical position and hardly notice it if at all. If the kinaesthetic elements they mentioned are not of this sort, I do not know what to make of them.

I have summarized the results of these inquiries in Table 7, which shows the number and percent of subjects in each group using various mental processes. These apply to sixty-one scientists only since I could not use the data from the other three. I have kept the experimental and theoretical physicists separate in these tables because they differ so sharply in this respect. By visual imagery I mean thinking in terms of pictures. These may be like memory images of real things, they may be diagrammatic, they may be symbols, such as mathematical symbols, but which are seen rather than said to oneself. And they may be in two or three dimensions, and manipulable or fixed. Auditory-verbal imagery means thinking in terms of words, as though you are talking to yourself. You don't actually say the words but you hear them in your mind as though they were being said. Thinking in symbols can also be classified here if you say the symbols to yourself rather than seeing them. Imageless thought is thinking that cannot be described in terms of any sensory modality, a sort of feeling of relationships, of "just knowing" something. Kinaesthetic thinking involves images of muscular movement of some sort, or at least of tension.

Each subject may appear in several columns; I have recorded all forms which have any significance for him (but not all he can use if he tries). Those who use verbal imagery only in preparation for communication are not included under the heading of verbal imagery. Failure to be included under any heading does not mean that the subject cannot use that form of thinking, but only that he does not usually use it.

You will see at once that the biologists are concentrated in the visual imagery group and so are the experimental physicists. The theoretical physicists are more likely to use verbal or other forms of symbolic thinking such as mathematical formulae, or only imageless thought. Over half of the total group use imageless thinking and I suspect that this is an underestimate, as noted above.

If you put those who use any form of visual imagery, with or without imageless thinking, but do not use verbal imagery, in one group, and those who use verbal imagery with or without imageless thinking but do not use visual imagery, you have a distribution like that shown in Table 8A. In this table each man appears only once. The subjects not included in this table either report imageless thinking or both visual and verbal in about equal amounts. It is possible to test statistically whether this distribution is one that you would be likely to get by chance if there was no relationship between being a particular kind of scientist and having a particular sort of thinking process. If you examined a lot of such groups and there was no relation between these two things, you would get, *on the average,* a distribution like that shown in Table 8B. The technique of finding out how often, in a hundred times, you would be likely to get such a distribution as 8A by chance is known as the chi-square test. By application of it we find that there is less than one chance in a hundred that

Table 7

Mental Processes Used by Subjects

| | Visual | | | | | | Auditory-Verbal | | | | Imageless Thought | | Kinesthetic | |
| | CONCRETE, USUALLY 3-DIMENSIONAL | | DIAGRAMS, GEOMETRICAL, ETC. | | SYMBOLS, VISUALIZED | | FORMULAE, ETC., VERBALIZED | | VERBAL IMAGERY | | | | | |
	N	%	N	%	N	%	N	%	N	%	N	%	N	%
Biologists	11	55	3	15	3	15	0	0	6	30	7	35	0	0
Exp. Physicists	7	78	3	33	3	33	1	11	0	0	6	67	0	0
Theor. Physicists	3	27	2	18	2	18	3	27	4	36	6	55	0	0
Social Scientists	3	14	0	0	0	0	1	5	11	52	15	72	4	19

Table 8

Type of Imagery Chiefly Used and Scientific Field

A. In This Group

	VISUAL	VERBAL	TOTAL
Biologists	10	4	14
Exper. Physicists	6	0	6
Theor. Physicists	3	4	7
Social Scientists	2	11	13
Total	21	19	40

B. Theoretical Distribution

	VISUAL	VERBAL	TOTAL
Biologists	7	7	14
Exper. Physicists	3	3	6
Theor. Physicists	4	3	7
Social Scientists	7	6	13
Total	21	19	40

you would get the distribution shown in Table 8A, if the two categories are not related. It would seem, therefore, that there is some relation between imagery and profession. This test, of course, tells you only that the two things are probably related; it does not tell you anything about the nature of the relation. It could be that possession of one type of thinking process predisposes to selection of certain kinds of vocations, or it could equally well be that working in one vocation tends to develop certain ways of thinking about things. Which of these interpretations is correct, or whether neither is, could be checked by studies of younger men in the field, and by studies of the development of this process. Here it is worth noting that the few biologists who are unlike the rest of their group in the way they think are usually unlike them also in

148

such things as major interests in school days, and the same is true of the physical and social scientists.

The problem of how these differences in use of imagery develop is one on which I have no definite information but I did discover a rather interesting relation. As I was looking over these tables and thinking about the position of the different men in them, I suddenly noticed that among the men whose imagery is predominantly verbal were several whose fathers were ministers, and even more whose fathers were lawyers. I quickly checked the rest, and the results are shown in Table 9. Here I have classed as verbal professions the law, the ministry, teaching, editing and so on, and as non-verbal professions medicine, engineering, pharmacy, and so on. Some of the fathers' professions could not be definitely classed in either group and these were omitted from the tabulations.

Table 9

Nature of Work of Professional Fathers and Imagery of Sons

NATURE OF FATHER'S PROFESSION	IMAGERY OF SON		
	VISUAL	VERBAL	TOTAL
Verbal	5	10	15
Non-Verbal	8	2	10
Total	13	12	25

This table, too, shows a significant association. That is, subjects whose fathers were in professions obviously requiring verbal facility are more likely themselves to be dependent upon verbal imagery than subjects whose fathers were in other types of professions. Again, the nature of the association is not known. It could be due to a genetic character, or it could be due to growing up with a father who, one might guess, did a lot of talking. The mother's attitudes were probably even more important if the association is due to environmental influence, but I have no information about them. Maybe talkative men marry talkative wives?

I thought it would be fun to look up some of the test data, and see if men with certain types of imagery did better on some tests than on others. I checked a number of items but will mention only those that showed important differences. On the VSM the verbalists, as you would expect, were better on the verbal test (but the difference was not certainly significant by statistical test; you could get that much difference in these small groups by chance, about 8 times in a hundred). The verbalists did just a little less well on the mathematical test than the visualists and the spatial test showed no difference.

On the Rorschach the total number of responses was markedly different. The visualist averaged 28 responses and the verbalists 62. The verbalists also responded much more rapidly than the visualists did. There are a few other differences on the Rorschach but the great difference in total number of responses makes these difficult to assess. I did not find any differences on the TAT.

In summary, then, we do find that there are differences in the way these scientists think which seem to have relevance to the type of science they go into, and which are related to the kind of profession followed by their fathers, and to certain test performances. These groups are small and these data are not very good, as I mentioned at the beginning of the chapter, yet the possibility of such relationships has important implications and I hope that the whole situation will now be thoroughly investigated.

I do not think it would be a good idea, if a valid test for imagery is devised, to use it for vocational guidance. That is, I do not think that a boy whose imagery is not of the type usual in one scientific group should be advised not to enter that field on that ground alone. It may be that, just because his mind does work differently from most in the field, he will make a very special contribution.

XII

How Smart Are Scientists?

How SMART are these eminent scientists? Very. This will not surprise you. On the other hand, what is surprising is how very wide the range is in this group, so far as their performance on the intelligence test I used is any indication.

When intelligence tests were first being developed, it was the general opinion that there was a particular capacity, called general intelligence. This was defined in various ways: "the ability to think in abstract terms"; "the ability to reason well or form sound judgments"; "the ability to utilize previous experience in meeting new situations."

Since intelligence developed with age, so that you could do at age 12 a number of things you could not do at age 6, a system for expressing the degree of intelligence as related to age was devised. This was the intelligence quotient, or IQ. This figure is a simple one. You get it by dividing the Mental Age (which is computed from the number and type of tests the subject could do) by the actual Chronological Age. If an 8-year old child, by passing some tests and failing others, scored a Mental Age of 8 years, his IQ would be $\frac{8}{8}$ or 100.

Similarly if he scored a Mental Age of 6 years his IQ would be 75, or if he scored a Mental Age of 10 years his IQ would be 125. As testing practice developed, tests came to be constructed so that about two-thirds of the population got IQ's of between 85 and 115, and in general the distribution of intelligence test scores follows what we call a normal curve. That is, most of the people are around the average, and the farther away you go from the average in either direction, the fewer people there are. This is shown graphically in Figure 1.

The IQ is actually not a very useful way of recording the results for adults. This is because intellectual ability does not increase noticeably after certain ages (the peak age differs with different types of intellectual ability), and declines at different rates of speed. In computing IQ's for adults, therefore, we do not use the actual Chronological Age of the adult, but a conventional age, now usually set at 14 or 16. If, however, one thinks of the IQ in distributional terms, as in the Figure, it has meaningfulness at any age. That is, an IQ of 100 always means that the possessor is at the average for that capacity for his age group. An IQ of 115 would mean that only about one-sixth of the group are brighter than he, and one of 130 would mean that only about one-twelfth of the group are brighter than he.

Theoretically the lowest IQ you could have would be 0, but then you would hardly be alive. We do not know what the highest possible IQ is. For one thing some people are smarter than our tests can test. When Dr. Terman and his associates were hunting for children with very high IQ's, the highest they recorded was 200. When Dr. Cox made an intensive study of the biographies of famous men she and her associates estimated their IQ's on the basis of reports of what they

153

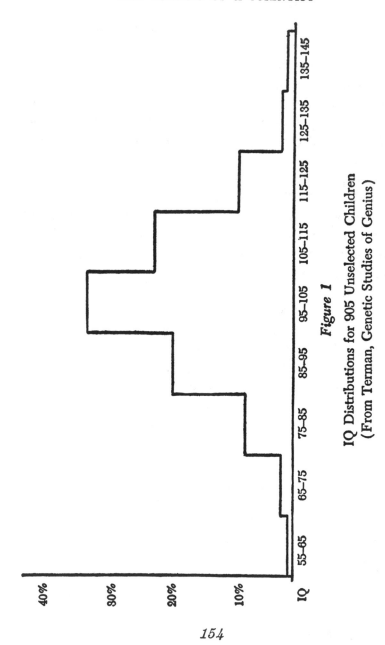

Figure 1

IQ Distributions for 905 Unselected Children
(From Terman, Genetic Studies of Genius)

could do at various ages. Here I have listed the scientists in her group with their estimated IQ's. Some of them you may not place immediately, but you will know most of them; and I have grouped them by their major fields, although a number worked in several.

MATHEMATICS, ASTRONOMY		PHYSICS, CHEMISTRY		PHYSIOLOGY, ANATOMY	
Arago	180	Berzelius	160	Bichat	175
Bailly	180	Boyle	160	Cuvier	175
Cardan	175	Davy	185	von Haller	190
Copernicus	160	Faraday	170	Harvey	180
d'Alembert	185	Gassendi	185	Hunter	160
Galileo	185	Gay-Lussac	175	Jenner	165
Herschel	165	Lavoisier	170		
Huygens	175	Liebig	180	NAVIGATOR AND EXPLORER	
Kepler	175	Priestley	165		
Lagrange	185	Watt	165	Cook	160
Laplace	190			Franklin, J.	160
Napier	170	BOTANISTS			
Newton	190	Boerhaave	165		
Pascal	195	Candolle	170		
		Linnaeus	165		
NATURALISTS					
Agassiz	175				
Buffon	160				
Darwin	165				
Humboldt	185				

They gave their two highest ratings to Leibnitz (205) and to Goethe (210).

It was assumed for some time that a person's IQ was a fixed part of him, like complexion or eye-color, and that, except in extraordinary instances, it did not change. This was what was called the constancy of the IQ. The term was first used by Terman and although he pointed out at that time that about half of the children he reexamined had shown changes in IQ (from 1 to over 20 points) this tended to be overlooked.

Our ideas on the nature of intelligence and on the constancy of the IQ have changed. There is, however, more agreement on the latter point than the former. We know now that the IQ is not constant in the sense we used to think of it, but that there are many things that may affect it, and that particularly in the very early years, we cannot effectively predict what any individual's IQ will be 10 years later. On the other hand, by the age of 7 or 8 we can get about as good an estimate as we are ever likely to, but we cannot be sure that environmental or emotional influences will not alter it to a greater or lesser extent. Shifts after that time, however, are under most circumstances sufficiently small that the measurement of intelligence is a very useful technique.

The nature of intelligence is a subject on which there is great and violent disagreement among my colleagues, but they do agree pretty well on some points and these are rather important ones. It seems to be generally accepted that there is more than one kind of intellectual capacity. There is still disagreement as to the hierarchical relations, e.g. is there one general factor plus a lot of special abilities, such as mathematical abilities, or are there an undetermined number of special abilities, all of equal rank? This is an oversimplification of the argument, but it will give the general idea. Among the different theorists there does seem to be agreement that there is at least one ability which can be called verbal ability, and this has played a very important part in the assessment of general intelligence as it has been made by most of the standard tests. Verbal ability means ability to understand and use language. One reason it has been so prominent is the association between testing and education. This ability is obviously of prime importance in making good grades in many courses in school. It has also played so important a role in test development, in my opinion, because as will appear, it happens to be an ability which psychologists have in large measure (or people who have it sometimes become psychologists). There is also fairly general agreement that there are special abilities involving perception of spatial relationships and ability to manipulate numbers. Beyond this, every worker in the field, practically, has his pet list of factors.* It is an important, and demonstrable, further point, that these various abilities are not closely related. That is, a man who is high in one is not necessarily high in another. Nor is he necessarily low in the second if high in the first. In statistical terms, there is a low correlation between them.

I have gone into these details to explain why I have used a Verbal-Spatial-Mathematical test, and why I have not combined the scores from all of them into one final, overall score. The test I used is not one that has been used before, at least in this form. When I was planning the study I looked over a variety of standardized tests. (A standardized test is one that has been tried on a population of known characteristics, so that we know what scores to expect on the average, and

---

* On top of all this, the most recent work is inclining to the view that it has been a mistake to try to separate intellectual behavior so strictly and to try to test it as a substantive unit. It is now being considered whether we might not do better to work at the problem in a different way and try to include other factors such as motivation. I strongly endorse this. For a long time I have been convinced that there is no such thing as "creative ability" as a unit factor which some people have and some do not, and I strongly suspect that I will soon come to the conclusion that the same thing applies to intellectual ability as a thing apart.

that has been checked to be sure that it is valid, that is, that it tests what it is supposed to test, and that it is reliable, that is, that you will get the same results if you give it to the same population again. These are difficult technical problems to which a great deal of attention is given.) I could find none that seemed to me to be difficult enough for the group I proposed to test. In psychological jargon, they did not have enough ceiling.

I took my problem to the Educational Testing Service, which, among other things, developed and administers the Scholastic Aptitude Test which many of you have taken as a preliminary to college entrance. After some consultation they pulled out a lot of difficult items from their files and made up the verbal test. The spatial test is part of another test, and the mathematical test is an abbreviation of a special test they constructed for one of the military services during the war. All were given with arbitrarily set time limits.

In testing you must consider whether your test is a "speed" test or a "power" test. In a speed test you are finding out how many items of given degrees of difficulty the subject can do in a set time,—that is, in part, a measure of how easily he can do these problems. In a power test you are finding out what is the most difficult item the subject can do in any amount of time. For my purposes I should have preferred power tests, but this was practically impossible. The verbal test as first tried out was not difficult enough for my first subjects so another section had to be added. It is a speed test for most of these subjects,—although, and this is another of those things that make life so difficult for psychologists,—for some of them it is surely a power test. The spatial and mathematical tests are chiefly power tests even though they were also given with a time limit. Few of the subjects could have done significantly more with longer time. As it turned out I was so concerned over getting tests with enough ceiling, I almost failed to get a floor under them.

I was not particularly concerned at the outset over the fact that I had no norms for this test. That is, I had no idea what any other population would do on the same test. I just assumed that eminent scientists were extremely bright people, and I did not particularly care just how bright they were. What I wanted to know was whether there was a pattern in the relative standing on these tests for any group, and if so, how these patterns compared. That is, I wanted to know if one group of scientists tended to be relatively high on one test and relatively low on another. The tests, although of different factors, and with different numbers of items, could be compared directly for any person or group, by converting the raw score (in this case the number of items answered correctly) into what is known as a standard score. The name refers to the standard deviation, a statistical measure which is used in computing the score. This score gives you the position of the subject with respect to the average and distribution of all of the scores in his group. If his standard score is 0, it means that he scores exactly at the average of the group; if his standard score is −.5 it means that he is at such a position below the average of the group that only one third of the group got a lower score; if his standard score is +1.0 it means that only one-sixth of the group scored higher than he.*

* This assumes a normal distribution of scores on the test, which will be near enough the case on this type of test with a large group. The formula for the standard score is $\dfrac{\text{Mean—score of individual}}{\text{Standard Deviation}}$.

With this system, then, one can compare scores from test to test directly.

I made an attempt to get some graduate students to take the same test, just as a matter of general interest, but succeeded in getting only 10, and under circumstances which made it impossible to judge how they had been selected. I then dropped the idea of getting any comparison group until it was announced that tests were to be devised in connection with draft deferment. It then seemed to me that I should make some effort to get a reasonably exact idea of just where these eminent scientists stood in the general distribution of intelligence in the population. Such information might be of use,—particularly in determining the level at which exemption might be considered on this basis,—and was not obtainable anywhere else. I had the great good fortune about then to meet an old acquaintance, Dr. Irving Lorge, who came to my rescue and arranged to give the test to all students matriculating at Teachers College, Columbia for a Ph.D. that February. All of their Ph.D. students have to take a battery of tests. This test would be included in the battery. Since the other tests had been well standardized it would then be possible to draw up tables of equivalents by which scores on the VSM could be converted (within certain limits of assurance) to scores on these other tests. This, incidentally, upset my budget considerably.

THE VERBAL TEST

This test, as originally used, consisted of 50 items, but a second section of 30 items was added. The task in all of the items was to select from a number of words two of which had opposite meanings. In each item in the first section, four words were given, and the subject had to pick the two which were most nearly opposite in meaning and underline them. Here is one of the items:

1. Predictable 2. Precarious 3. Stable 4. Laborious. Which two words would you underline? (The correct answer is 2 and 3.)

In the second section the task was the same but it was presented a little differently. This time, one of the opposites was given, and the task was to pick the one of 5 other words which was most nearly opposite to the first one. Here is an example of that group:

ABSOLUTE; 1-forget 2-usurp 3-absolve 4-utilize 5 limit. Which word did you choose? The correct one is 5.

All in all, then, there were 80 items to start with, but one of these had to be dropped. This is because so many of the biologists picked the same "incorrect" answer, about as many as picked the "correct" one. It seemed to me that as good an argument could be put up for one answer as for the other, and as the Educational Testing Service agreed this item was not scored. It had to be left in the test for later groups, since those of the first group who had reached it had had to spend some time on it. If it had been omitted for the later groups the time allowance would not have been exactly the same in all groups. Fifteen minutes was allowed for this section.

The average scores and ranges of scores for these groups are given in Table 10. These are the number of items done correctly. I have kept the physical and social scientists subdivided. The best man on this test missed only 4 items, and a number failed fewer than 10. The test, then, still has not enough ceiling for this group.

161

Table 10

Scores on the Verbal Test

	AVERAGE	RANGE
Biologists	56.6	28–73
Experimental Physicists	46.6	8–71
Theoretical Physicists	64.2	52–75
Psychologists	57.7	23–73
Anthropologists	61.1	43–72

Let us look at the averages. The experimental physicists are the lowest, the theoretical physicists the highest, and the average for the psychologists is exactly that for the total group, 57.7. Clearly this is not an ability which is nearly so important for the experimental physicists as for the others, or else there is something funny about the test in their case.

It should be said that under some circumstances any test may not give a fair estimate of the capacity of the subject. He may not really have worked at it for some reason, he may be sufficiently ill that the test is affected, or he may be very depressed over something and this would tend to lower his score. These are situations which the clinician is supposed to be able to judge, and situations for which he always watches. It is one reason why it is usually better to administer the test in person than to leave it with the subject, although the latter course gives him an opportunity to work on the test when he is feeling like it.

The question is, then, are such circumstances operative in this situation, and to any great extent. At the time the test was given I noted that the biologist scoring lowest, and the experimental physicist scoring lowest may not have done justice to themselves. (The note was made before I had scored the test.) I have no such notes for any of the others and I do not think the other tests are in question. In the case of these two men, I felt that the tests were not seriously lower than they were able to do at the time the test was given, but that they could probably have done better at other times. On the other hand, it is impossible to be sure how much better they could have done,—I can only offer my own opinion that it would not have put them in very different positions in the group.

A way of getting around this, if the scores are used at all in comparisons, is to use the median scores instead of the average scores. The median score is the one made by the man who stands at the middle of the group if they are arranged in order from lowest scoring to highest scoring, and it is not affected by extreme scores at either end. I have done this in making estimates of the meaning of these scores in terms of other tests, but the differences are not very great. For the verbal test the average is 58 and the median is 62.

It happens that the six men in this total group who score under 40 on this test are two experimental physicists, two biochemists, and two psychologists, both doing physiological psychology. Four of these men were Phi Beta Kappa in college,—that is they made extremely good grades. (Of the other two, one probably did not make good grades, and I think there was no Phi Beta Kappa chapter at the college which the other attended.) But the specification of their particular fields suggests that they have gone into fields in which this particular ability is not too essential.

163

Let us see what these figures mean in other terms. I must caution that these equivalents have been arrived at by a series of statistical transformations based on assumptions which are generally valid for this type of material but which have not been specifically checked for these data. Nevertheless I believe that they are meaningful and a fair guide to what the situation is.

The median score of this group on this verbal test is approximately equivalent to an IQ of 166. The lowest and highest scores would be at about the levels of 121 IQ and 177 IQ. We have already noted that the test was probably not difficult enough for some of these men, hence this estimated upper limit is lower than a more adequate test would give. This would not affect the median score, however.

Now let us look at IQ's of college populations of today. Embree found that 1200 high school graduates who went to college had been found during childhood to have a median IQ of 118; those who graduated with a B.A. an IQ of 123. Honor graduates had a median IQ of 133 and those elected to Phi Beta Kappa of 137. The range of IQ's for all of those who received degrees was from 95 to 180. For persons who went on to take a Ph.D. Wrenn found a median IQ of 141.

It is clear, then, that so far as verbal ability is concerned these eminent scientists are on the average higher than the general run of those that get Ph.D.'s, but, and this is very important, some of them are not as high as the average Ph.D. It is, then, not essential to have this ability *at the highest level* in order to become an eminent scientist. That it is doubtless a great help is another matter, but it should be remembered that it is less helpful in some fields than in others.

THE SPATIAL TEST

This particular test was suggested by the Educational Testing Service because they considered it the "purest" spatial
test they had. That is, it seemed to be least related to other
tests, in the sense that performance on this test could not be
predicted on the basis of performance on any other test. The
results with it have been very interesting but rather unsatisfactory. For one thing, it does have a significant correlation
with age at these levels, and for another, I am not quite sure
what it is testing.

There are 24 items on this test, and the subject has 20 minutes in which to do them. He has, first, three practice items.
I have reproduced on the following page the instructions, and
these three items. How did you do on them?

The scores this group attained, in terms of number right,
are given in Table 11. Again the top scores are so close to the
number of items in the test that the true upper limit is probably above that recorded.

Table 11

Scores on the Spatial Test

	AVERAGE	RANGE
Biologists	9.4	3–20
Experimental Physicists	11.7	3–22
Theoretical Physicists	13.8	5–19
Psychologists	11.3	5–19
Anthropologists	8.2	3–15

The theoretical physicists are also the highest on this test,
but they are followed this time by the experimental physicists

Each set of four drawings in this section contains two views, and only two views, of one solid figure, taken from different angles. The other two drawings represent solid figures which differ in some particular from the one solid of which two views are shown. You are to identify the two drawings of the same solid figure that represent the same solid and encircle the corresponding numbers. You are to encircle two numbers for each problem.

PRACTICE PROBLEMS

Consider practice problem A. Figures 1 and 2 are not views of the same solid. If figure 1 were moved into the position of figure 2, the small prism-shaped block would not match that in figure 2. Likewise, blocks 1 and 3, and 1 and 4 are obviously different. Similarly, 2 and 3, and 2 and 4 differ. Block 3, however, is actually another view of the block in figure 4. Hence the answer to problem A is 3-4.

In problem B a careful comparison of the figures shows 1 and 3 to be different views of the same block.

In problem C the answer is 2-4.

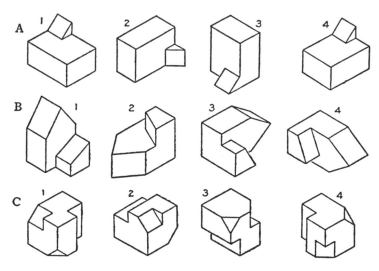

Figure 2

and then by the psychologists. The anthropologists are the lowest of the groups. The average for the total group is 10.9, and the median is 10.

These scores can also be translated, as were the scores on the verbal test, into approximate IQ terms. The lowest score, in these terms, is about the same as the lowest score on the verbal test; it gives an IQ equivalent of 123. The highest score on this test, however, does not quite reach the median score on the other, being equivalent to an IQ of 164, while the median IQ on this test is 137. This means that, as compared to Ph.D. students these men do not score as much higher on this test as they do on the verbal test. But there is a catch here. That is the correlation of this test with age which is —.40. That means that the younger the man the more likely he is to get a high score on this test. If these men had been tested 20 years earlier they might have scored as much higher on this test as they did on the other.

The theoretical physicists are the youngest group, so this may in part account for their higher average on this test, although only in part. It seems quite possible that this particular ability does have some relation to work in physics, but these data are only suggestive of that, and other data are needed to clarify this. We do know that in general arts students in college do less well on non-verbal types of tests than science students do.

THE MATHEMATICAL TEST

This test was taken from one which the Educational Testing Service had developed for a special project. The original was too long for my purposes, so we selected portions of it, omitting some of the easiest items, and then deleting other items of varying levels of difficulty. There were 39 items in the final form and 30 minutes was allowed for work on it. No subject finished the test. The items were generally of the type known as mathematical reasoning, and an example is given below:

Select the correct answer:

$$\text{If } x + 3y = 7x + 5y, \frac{x}{y} = (?)$$

$$(A) -3 \quad (B) -\frac{1}{3} \quad (C) -\frac{1}{9} \quad (D) 1 \quad (E) 3.$$

If you did it properly you underlined A.

This test was given only to the biologists and the social scientists. I tried it on a few of the physicists just to see. It bothered one of them, but the others sailed right through, making an occasional careless mistake. The test was obviously not difficult enough for them and a waste of their time. It would be clearly impossible to get a test which would be difficult enough for a first-rate physicist and still be easy enough for most biologists and social scientists. The data given in Table 12 are therefore only for the biologists and the social scientists. This test clearly was difficult enough for all in these groups.

Table 12

Scores on the Mathematical Test

	AVERAGE	RANGE
Biologists	16.8	6–27
Psychologists	15.6	8–27
Anthropologists	9.2	4–13

The biologists are a little better than the psychologists at this, but the difference is not enough to be significant,* and the anthropologists do not do particularly well.

Let us look at the equivalents on this test. It is not correlated with age (the correlation coefficient is .00). The lowest score on this test is about equivalent to an IQ of 128, the median score to an IQ of 154 and the highest to an IQ of 194. That is very high indeed. Mathematical ability is certainly important for work in physics, but it seems that it can also be important in some other sciences, particularly biology and psychology.

It is interesting to look into the kind of work done by the men with the highest scores. The two highest scores attained by biologists were made by geneticists. All of the geneticists average 21.9 on the test, whereas the other biologists average 14.0 on it. This is not surprising when you consider how much of modern genetics is mathematical. The highest scores made by psychologists were all made by experimentalists, but so were some of the lowest scores. On the average, though, the experimentalists do come out higher than the others (but there are only 4 of the others). The averages are 17.7 and 10.2.

COMPARISONS OF THE TESTS

Differences in level of ability are also shown if you look at the patterns of scores for each man. These can be generalized, although there are exceptions in each case. On the whole among the biologists, the geneticists and biochemists get

* Significant in this usage is a technical term. It means that you would not be likely to get such a result by chance in the size groups you have more often than 2 (or whatever figure you set, usually 2 or 5) times in a hundred. This is considered fairly safe.

higher standard scores on either of the non-verbal tests than they do on the verbal, and the other biologists reverse this, tending to be relatively better on the verbal. Among the physicists there is some tendency for the theorists to be relatively higher on the verbal and for the experimentalists to be relatively higher on the spatial. No special tendencies are noticeable among the anthropologists but among the psychologists all but one of the experimentalists has a higher standard score for either spatial or mathematical than for verbal.

If we examine the correlations between the tests we see that it is true in this instance that ability to do the mathematical test is not related to ability to do either of the other tests. The correlations are +.14 and +.21 which are not significant with these groups. There is, however, some correlation between the spatial and verbal tests. The coefficient is +.33, and that is high enough to indicate that some relationship exists. It is not close, but it points up one of the difficulties with the spatial test. That is, it can be done in different ways. Those who do it extremely well for the most part do it without much conscious reasoning about it. They can tell the answer "just by looking" at the figures and imagining them turned around in various ways. Some of the others, however, are able to do it fairly well by going through a fairly intricate process of reasoning it out by talking to themselves about it, and it is in such circumstances, I think, that the relation with the verbal test comes in.

It seems quite clear that, having chosen your general field, •the particular kind of work you do in it is related in some degree to your particular capacities. How well you do in the field is partly a function of your capacity for that particular field, but even more a function of how hard you work at it.

170

XIII

What Kind of Stories Do They Tell?

THIS CHAPTER refers to the Thematic Apperception Test, not to the excuses scientists give their wives when they stay late at the laboratory! (By now their wives are used to this and probably pay no attention.) This is a terrible name for a test (but after all it was developed at Harvard) and it is no wonder that it is always spoken of among psychologists as the TAT; let us call it that, too.

The TAT, then, is a test of personality. The task of the subject is to make up stories about pictures which are shown to him one at a time. Most of the pictures are not very good pictures, artistically speaking, as I was told at great length by the artists whom I studied. The artists spent so much time complaining bitterly about the pictures that it was very difficult to get them to tell me any stories about them. It was a great relief when I worked with scientists, most of whom were not so disturbed by this aspect, although some of them had some unfavorable comments to make, too. Several of the pictures are reproductions of modern painting. Even this did not help much with the artists, far from it in fact. I only used one reproduction with the artists, but of course they all recognized it. Unfortunately some of them did not like the artist who painted it, and indeed one of them became so incensed that I gave up the test altogether before he threw both me and the pictures out of the window. He is a bit on the violent side, anyway.

The TAT is one of the group of tests known as projective techniques. The Rorschach is another. There are certain aspects of personality, wishes, strivings, imaginings, and so on, some of which remain largely outside of the consciousness of the individual and yet which play an important role in the determination of his behavior, and his satisfactions. Furthermore even those which are conscious are not always easily communicable and often one does not see that they have any bearing on behavior. Therefore our first approach of personality investigation, which was by means of quite direct questionnaires, came to be supplemented by techniques which use somewhat more roundabout methods. Projective techniques are quite varied in the actual situation each utilizes, but all are alike in that the subject is presented with a relatively ambiguous situation and requested to do something with it. The possible responses are extremely numerous and the subject has a minimum of external guides, hence what he does is largely dependent upon himself, upon his own usual ways of looking at things, upon the things that come most readily to his mind, upon the forces that chiefly move him. Furthermore he usually has no idea of how the responses he gives are to be utilized in making inferences or even what sort of inferences are to be made from them. In fact, the intent of the test may be deliberately disguised. People who come to psychologists for help are often of two minds about it, wanting it badly and at the same time holding out on the psychologist stubbornly.

Clearly when a minimum of restrictions are placed upon the subject in his handling of the material offered him, there is maximum variability in the responses and consequently maximum difficulty in ordering these for purposes of scoring and interpretation. In the VSM, just discussed, the subject was limited to 5 or 6 possible answers, only one of which is correct. In projective techniques, on the other hand, not only are the possible answers very little limited, but there are no correct answers. Some are more often given than others, but any response the subject gives is "right" for him. Whether the experimenter will be right in interpreting the results of the test is another matter.

This is no small problem. We have spoken of validity and reliability in connection with the VSM, that is, of being sure that the test tests what you think it does, and that it would get the same results upon another occasion. If it is difficult to determine these with intelligence tests, as it is, think how much more difficult it is with projective tests. What criteria can be set up to determine the validity of a personality test? This is especially difficult with a test which is not broken down into a series of isolated elements, the validity of each of which can be determined separately. In most projective tests, although certain elements are considered separately, the interpretation of each depends upon the total of all of the elements and their interrelations. Perhaps the most useful technique for validation is some variation of a matching technique. For example, an experimenter gives the test to a number of subjects and writes personality descriptions of each. Then these are shuffled up and given to one or more judges who know the subjects, and who attempt to match up the description with the subject. Such experiments have been reasonably successful in a number of instances but they do not validate specific inferences from the tests. Other techniques have been tried, such as the success of the tests in differentiating between patients in different psychiatric categories. This is not very useful because of the fact that these categories are still not well defined and there is little consistency from one psychiatrist to another in using them, e.g. the same patient may be diagnosed manic depressive and schizophrenic by different psychiatrists at the same time.

On the question of reliability there are equal difficulties. In intelligence tests this is usually checked by one of several methods, of which the commonest are repetition of the test after an interval of time, and the split-half technique by which odd items are put in one section and even in another and the two sections correlated. Again, because of the nature of projective techniques, neither system is applicable and in fact no really adequate system for checking the reliability of any of the projective techniques has been devised.

There are two additional major problems. One is that the ordinary statistical methods are not sufficient for handling the complicated material of projective techniques. Some form of pattern analysis is probably required, but so far no completely satisfactory one has been devised. One of my colleagues, however, has just written me that he thinks he has this problem about licked.

Another problem is the experimenter. Variation between one experimenter and another, when both are well trained and check themselves often, is not great for individually administered intelligence tests. With tests like the VSM it is practically non-existent; you just get a scoring key and check

174

the answers, and within the limits of careless error anyone who scores the test will get the same results. This is an additional problem with projective techniques,—will two psychologists, looking at the same data, make the same inferences? In fact, experts will usually agree on the main outlines but there will almost certainly be differences in emphasis and in details. In these techniques the experience and the skill of the examiner are probably the most important factors in the usefulness of the test.

Why, then, in the face of all these difficulties and uncertainties, are these now among the most frequently used tests in the clinic? For the very simple reason that they *are* useful, that they give information which is of enormous help to the psychologist or the psychiatrist in understanding the difficulties of the patient and how they developed and in estimating his assets and liabilities. They are of equal value in research with normals.

It must also be stated that in the past few years the difficulties outlined briefly above, and others, have been studied extensively, and there is a perfect spate of research papers appearing on them. This is in marked contrast to the earlier days when the proponents rested smugly upon the clinical usefulness of the tests and the rigorously minded experimentalists would have nothing to do with them. We are learning a great deal about the tests. More important, fuller study of the tests is helping us in our attempt to work out some consistent theory of personality development and modification.

Use of a test as a clinical instrument in the study of one individual person, and use of the same test in a research study where you want to compare groups of persons on the same test are rather different problems. There is no particular difficulty when the test, or even some aspects of it, can be expressed mathematically, because average scores can easily be computed for each group and these averages compared as was done with the VSM. With the Rorschach something of this sort can be done, although it only touches upon a portion of the material available from the test. The TAT permits even less of this sort of mathematical treatment than the Rorschach, although with it, too, some of this can be done. It is important, though, not to be seduced by the ease of mathematical manipulation into forgetting the more important but more difficult other aspects of the tests.

The material for the TAT consists of a set of cards, on which are pictures in black and white. Some of the pictures are fairly sharp and clear and some are rather ambiguous. In all, the task of the subject is to tell a dramatic story about the picture. He is asked to tell what is going on at the time of the picture, what were the events leading up to this and what the outcome will be, and he is also asked to tell what the characters pictured are thinking and feeling.

In the standard administration of the TAT it is usually stated that this is a test of imagination but I did not follow this. There are good reasons in the clinical situation for sometimes concealing the nature of the test. In this situation it seemed to me unethical to do so. Therefore I had explained in advance to my subjects that the TAT and the Rorschach were designed to show aspects of personality functioning. I did not elaborate upon this nor did I explain in advance how they do this, but this variation from the usual technique must be noted.

I also altered the standard technique in another way. The

standard TAT uses 20 cards, and is administered in two sessions of 10 cards each. I could not give this much time to one test in my set-up and therefore I did not follow this either. When I first began the work with individual subjects I had it in mind to use certain of the cards regularly and to add others as seemed indicated for the individual case. (Some of the cards are more useful than others for getting at certain things.) I shortly settled onto a regular set, however, as the other system would make it much too difficult to handle the data for the group. There are partially differing sets for boys and girls, for men and women. I used Cards 1, 2, 4, 6, 7, 10, 13, 15, and 11 from the set for men, series of 1943.

I cannot reproduce any of the pictures for you. If the material used by psychologists is given wide circulation, then if you happen to need a psychologist some day you and he will be hampered by the fact that you may not be able to react spontaneously to these materials. A casual glance at the material would probably make little difference, of course, but the amount of difference it made would vary from one person to another, and it would be hard to tell what it was in the individual case.

As an example, however, let us look at the photograph opposite. Suppose that this was a TAT card and you were asked to tell a story about it, about what is going on now, and what had gone on and what will go on. What would you say? You can write your story on page 179 which has been left blank for you.

Here is what several other persons have said.

My story is:

"These boys are working on a paper. The one on the left is trying to persuade the others that they should do it in a particular way, but the one next to him is just waiting for him to finish talking so he can put another side of the case. The others are really listening and will think about it. I don't know what they will decide to do but I suspect that the boy on the left will convince the open minded ones."

(I was interested to note, after I heard the other stories, that there is something of a conflict over dominance in mine. I had thought that this was one way in which I was now not like many other psychologists, but now I wonder!)

My husband said:

"Well, in this picture we have a group of 4 boys who are members of a boys' club and the leader, a school teacher. He's interested in boys and finds teaching school a thwarting experience. Since he likes boys and wants to broaden his experience with them he has formed this club which carries over outside of school and makes him a better school teacher. He's been working with them on a project which consists of taking press hand outs and analyzing them for propaganda and trying to rewrite them, slanting them from this angle to show how it would be written with these axes to grind. The purpose, of course, is to teach the boys to recognize bias and to discount it. One reason why he is dissatisfied with the strictly school relationship is that he can't do this in school. For instance, he certainly wants the boys to recognize the communist bias so he has assigned to one of them the rewriting of this press hand out from the communist point of view. If he did this in school he would probably be accused of being a communist himself. The boys are tremendously interested in this task and are getting a great kick out of it. They are interacting in a very thoughtful way with each other and with the teacher of whom they are fond. These boys will

WRITE YOUR STORY ON THIS PAGE

shortly be voting citizens and even more shortly will be in the army and I hope that they are not too disillusioned when they find out that not all people are as thoughtful and as honest as their teacher."

(It has doubtless occurred to you that this story was told while the threat to the universities from Congressional committees is in all of the papers.)

My publisher says:

"Well, I guess a bunch of kids are studying one of those school newspapers that I haven't very much use for and probably it's a civic events class. Either he's reciting it and the others are comparing their own memories and impressions or else they are studying for it. I guess reciting because the man on the left of him looks like a teacher. It looks a little bit to me like an ad for that sort of newspaper, that it's posed rather than genuine. What's going to happen next? Well if it's posed they are just all going to get up and collect their fees and go home. It looks too much like that to me to get anything real out of it and if I tried to imagine a story with any sincerity to it I'd just have to make it up."

(There are several fascinating touches to this, as you will judge later.)

But what has this to do with the kind of person the narrator is? Actually this test gives a good deal of information about the particular strivings and sentiments and attitudes of the person and sometimes something about how these were developed. For the formal analysis of a series of stories of one individual, there are a number of schemes. Details of these are intricate and rather burdensome. It is important to note such things as how closely the subject sticks to matter-of-fact statements, whether he includes what the characters are thinking and feeling, whether he makes the stories fairy-stories or movies, or tells them as actual events, whether he includes past and future as well as present, whether outcomes are favorable or not, whether the general tone is cheerful or sad or sardonic or what and which of his stories are unusual ones. It is important, too, to notice how he differs in these things and others from card to card. Then, too, one must note his attitudes about human relations, about social institutions and so on, as these are reflected in the stories and the kind of language he uses about them.

With all of these and other items recorded story by story, analysis is still far from simple. It has been shown that the sources of stories told by subjects include: a. books and mov ies; b. events in which an acquaintance has participated; c. the subject's own experience, objective and subjective; and d. conscious and unconscious fantasies. It is clearly not easy to disentangle all of these possible sources and to say which ones are personally important for the subject. It is relatively easy in a story in which one character is the major protagonist, and the subject appears to have identified with this character. Even then there is the problem of deciding whether what happens to this protagonist is what the subject wishes or fears will happen to him. Only considerable experience with the test and keen clinical intuition can get around these difficulties, particularly when the test is being analyzed "blind" that is, when the analyst has no information about the subject other than age and sex. If, however, in addition to the test there is other material, either other tests or biographical data, then it is clear that one acts as a check on and as an illuminant of the other and the usefulness of the TAT is much greater. This is the ideal situation and this was available in the research reported here.

Before going on to consider what general statements can be made about these groups, let me give a few specimen stories to the same picture, and a few comments on them. Remember that each subject looked at the same picture so that the great differences in the responses are due to differences of some sort in the subject themselves.

The card is number 2 in the series, and depicts a farm scene, with a man plowing in the background, a woman leaning against a tree and a girl carrying books in the foreground. The most usual interpretation refers to a family, but note here the differences in decision as to the relationships of the three figures: father, mother, and daughter; a triangle; a mother and son and the local schoolteacher; mother, son and daughter; two sisters and a brother. Some of the subjects decide that the farm is prosperous and some that it is not. Some read conflict into the situation and others do not, and the degree of elaboration of the story is also extremely varied. The first three stories were given by biologists, the next three by physicists and the last three by social scientists.

Story 1. "This is a story about a peasant village somewhere, with an obviously European gal in front, a local gal that somehow has got educated and has worked up to be a schoolmarm. The folks in the background are not hers but a mother and son. The gal is in love with the fellow driving a horse but if she goes in for that she will be stuck on the farm like the other woman and she doesn't know what she will do. This doesn't explain incidentally why the lady is standing there with her hands on her stomach and not going to work. The gentleman isn't very interested, or maybe he was up to this moment, and he just has his back turned to demonstrate he isn't interested any more. I don't know what she will do because she doesn't know.

Comment: The career versus marriage conflict is not uncommon in the stories given by this group and it usually happens that the career is selected. These men value an intellectual life highly on the whole. But this subject cannot decide what the girl will decide (because she doesn't know,— a strong suggestion that he has identified with her) and he is uncertain, too, about the feelings of the man, who will not face the situation directly. There is a good deal of uncertainty about feelings in his whole protocol (this is what we call the full test record) and considerable evidence of aggression masked as passivity.

Story 2. This is a kind of rough one for me to figure out, but I guess this is a farmer who has been plowing his field or disking or making furrows. He's almost finished. It's along towards afternoon, late afternoon and his wife is waiting for him to finish. This girl had me stumped for a while but I guess she's a daughter and is on her way home from school with her books. And I guess the wife is patiently waiting for him to finish so she can walk back with him. She has everything all ready so she doesn't have to cook any more, and has come out to be with him. As a matter of fact I guess he isn't plowing, he's cultivating, these are rows of plants. I guess there is more to it, too, it looks like they have had an argument or something, the girl wants to do something but she's been overruled. It's been between her and her mother, I don't know

183

what part the father had. I guess that's all I can do with it. The mother is very definitely being a martyr about the whole thing.

Comment: There are several interesting things here. Note that the wife has arranged household matters so as to spend more time with her husband,—one would take it that this is approved of,—but there is a decided shift in attitude toward the same character in her maternal role. Note, too, the frequent use of "I guess" but the fact that he usually comes to a decision about the point, until it comes to paternal attitudes.

Story 3. Well it's evidently a farmer and his wife and daughter. A pretty primitive sort of a farm. The girl, I judge, is on her way to school. I can't quite make out the attitude of the woman, her expression is beyond me. I don't know what she has got on her mind. I should judge that one event that is going to happen is an addition to the family. I suppose that one could make up some ideas from the sorrowful expression on the girl's face. It doesn't look like a very prosperous sort of place, I suppose she thinks she has a hard time of it. I wouldn't say the future looked very hopeful. I suppose they will go on as they are, because the farm doesn't look like a very prosperous one.

Comment: Like the first two, this is a rather unadorned response, and without any extension in time except on a very general basis. Again there is uncertainty about attitudes, and the phrasing, "she thinks she has a hard time of it" calls for attention. The first two subjects did not notice, or did not mention, that the older woman looks pregnant.

Story 4. It's really a bad proposition. I've seen a lot of art exhibits and I wonder about what the artist had in mind. Let's see. This girl seems unhappy about, presumably, a discussion with her mother relating to the importance of working on the farm as compared to reading books on her school work. She seems disgusted with the attitude of her mother and her prospect of being stuck on a farm. (He stopped here and I asked about the future.) Oh, she'll break away some time and decide to either get a job in the city or go to school herself. I don't know whether she'll live happily ever after or not.

Comment: This story is brief but actually much more definite than the others given so far, and he is quite definite about the future in one sense, that the girl will break away. Happiness is a different matter. Note, too, the acceptance of the girl's attitude towards her mother, and the complete omission of the male figure in the picture.

Story 5. I can't locate the country, it doesn't look like the United States. It's probably unimportant. It's probably Europe and I should say this is quite clearly the family, a peasant or small farmer and the daughter who is going to a modern school. The mother looks quite content with her lot in life whereas the girl looks quite discontented and perhaps ambitious to have some other future than that of a small peasant farmer, and she sees this in the books that she's carrying. She's going to attain more freedom through education and so on. Also she looks as though she might be quite capable of doing it. Perhaps not a completely pleasant character, perhaps a little bit more ambitious than is consonant with being a pleasant person. She might be a little bit hard on her husband. She is probably very capable though. (Again I had to ask about the future.) She's going to succeed, she's going to get the education and she's going to get a different life than the one that's depicted here.

Comment: This story has some similarities with the last in the omission of the man except in passing and the statement of the girl's attitude and probable success. It, too, is qualified, but in a different way,—not with regard to the girl's happiness, but with regard to that of her husband, an unusual and interesting touch.

Story 6. Oh, this is in France, it's a farming family. The trouble is that here I'm going to be repeating a movie I saw rather than making up a story. The girl on the left is going off to school. Her big brother is doing the plowing. This must be her aunt loafing off to the side, she ought to be working somewhere. The central character is the school girl here, she's a nice kid. It's still a group picture and you feel that all the characters fit into the scheme of things on this farm. The farm is doing well and things are busy and her brother is at work and in the background her father is driving a team and it isn't harvest time yet. As a matter of fact the crops are still young, but it must be summer. (I asked for the future when he stopped here.) In the future this girl is going to get married. I might have to reconsider this other lady here. As a matter of fact she's to become a mother and I think we'll have to say the lady on the right is the mother of the girl on the left. It's a prosperous family and they are doing well now and since they apparently are hard workers they should continue to do well.

Comment: There is more description here than story. He

even notes the small background figure which the others have
ignored, and designates it as the father. Note the shifting of
the relationships, particularly from aunt to mother. Note, also,
the implied causation, hard work will bring success.

Story 7. This is a triangle. That's the farmer who is, who
is having this affair with this girl, and his wife is very sus-
picious and she's out there watching to see what goes on.
They are just at the point where the thing is just about ready
to break, but everybody is still sort of friendly. The man is
sort of ignoring the situation. The wife is keeping a close
watch on everybody and the girl is sort of looking off, she
doesn't know quite what to do about it all. Well, the future
will be that probably not very much will come of it. The whole
thing will blow over and the farm will go on. This complication is that the wife seems to be pregnant.

Comment: Other subjects also interpreted this as a triangle
situation, and the rather acceptant attitude is not unusual,
but the fact that it all comes to naught, and that the future
is expressed in terms of the farm, not the people, is very individual. That the man ignores the situation and goes on with
his work is quite characteristic of this group in general.

Story 8. Well, here we have a peasant family, and one, the
the girl at the left, is very much torn. She's in love physically,
emotionally and so on with the man who's cultivating or
something in the field, but she's an intelligent girl. She has
her books. She was picked out by the village priest as being
bright and destined for a teacher or something of that sort.
So she's dressed differently from the girl on the right who
has accepted the emotional definition of herself provided by

her culture and is at peace because, in fact, she's already pregnant slightly and from the man who is down there. She's feeling a certain triumph. The girl isn't dressed like a peasant. She looks like one but she's broken away from peasant customs. The girl on the right is thinking, 'Well you have got your books and learning. You will be in a better social status than I but I'm happier. I've got the man we both wanted. I've got a child by him. I'll be here. This will be our land and I'll bring him his lunch as I do now and while there will be a part of you he will always miss this will be a good life.' The girl at the left is thinking something of the converse of these thoughts. She's thinking, 'I'm doing well in my education. I'm interested in it. I refuse to give it up and yet I wonder if it's going to bring me any happiness. I can't help being a little jealous of my cousin there who's going to be a peasant all her life. But this is what I must do.' The man himself isn't thinking at all. All of this is in terms of the unconscious. In terms of the unconscious he's deliberately turned his back on this girl. He's accepted the situation. He has content enough with his wife who's a good hard worker, dependable and not unsatisfactory sexually, who's responsive, loves him and so on. Perhaps she is less inhibited than the girl. He's not thinking all that. He's just adjusted to his wife's bringing the lunch. The work is done. He's plowed a good furrow, but in the unconscious there is a feeling of partial fulfilment and that is why he has turned his back. (He, too, was asked about the future.) I forgot about your wanting the future. I didn't foresee anything.

Comment: Here is a fuller explication of the career versus marriage conflict which is presented baldly in many of the other stories. Note how much more fully he is concerned with the feelings of the characters, including those of the man, although, as often happens, the man's position in the picture with his back turned, is interpreted as symbolic of his deliberately turning his back on a situation.

Story 9. In the first place these aren't real people. I have to make them real to tell a story about them, but they are not real. They are stiffer than Grant Wood people and that's pretty stiff. It isn't a real picture. Women don't stand like that and people don't cut furrows as straight as that, particularly with horsedrawn plows. These people are two sisters and a brother. They have lost both parents and have jointly inherited a farm that previously belonged to the parents. The son is happy to remain on the farm and till the soil and he's going at it. He has a modern technical education in farming and agriculture. He has great ambitions to do things with the farm that he could not do when his father was alive, because the old man had definite ideas of how a farm ought to be run and wasn't interested in new-fangled notions. Some of the things he wants to do are crop rotation, modernizing his equipment, getting tractors and that sort of thing. One of the sisters, the one leaning against the tree, has decided to stay on the farm until she marries. She is going to get married pretty soon, we hope, because she is pregnant. The father of her child is the son of a farmer who lives nearby and the wedding date is set. In this particular region there is nothing especially wrong in a pre-marital pregnancy followed by marriage. The big thing is to get married before the advent of the child. This woman is pretty far along, too. The third sister wants to leave the farm. She has wanted to for a long time but couldn't because her parents didn't wish it. They didn't force her to stay on the farm but she knew it would upset them so much that she did not go away. But she is glad now that she stayed and also that she can now continue her education at the country school. As a matter of fact she has been away at school and is back now on vacation. She has brought some books she wanted to study. Her ambitions are in the direction of being a famous poetess or author. She's a dreamy type of person, very susceptible to emotional conditions in contrast to her sister who is utterly practical. This girl whose name is Mary is about 22 and continues her college education in the future and gets her degree, but she will never be a famous poetess or author. She will instead get married and have a family. The other sister is due for a bad surprise because her fiancé will not live long enough to marry her so she will continue to live on the old homestead. Her brother who is an understanding sort of person and helpful will support the child. For a number of years this woman will remain unmarried and will be in bad reputation around the neighborhood because of her illegitimate offspring. The boy, too, it will be a son, will be ostracized within limits. But two good things will come out of it. One is that this woman will eventually, simply by her conduct, be reaccepted into the community as the memory of her disapproved behavior grows dim, and she will eventually marry again. She will marry a young widower who comes into the region and buys a farm. He courts her in standard fashion and learns of the child from other people, but he doesn't mind and wants to marry her and adopt the boy. The boy will be prevented from a negative attitude of his contemporaries by the attitude of his mother and the fact that his uncle loves him dearly and treats him as his own son. So that the unhappy facts of his origin are overcome. The son in this picture will never become married but he will be an important person in his social group because he will gradually introduce modern and more efficient methods of farming, almost exclusively by setting an example, by showing that using hybrid corn means getting a better crop, and that there are genuine economies in laying out money by getting better machinery and so on.

Comment: (No, this subject was not raised on a farm.) Such stories as this last, with their elaborate and detailed development of the future and the quick departure from the exact moment pictured are relatively rare among this group. Note how far this subject goes beyond the material given him. This happened only with one biologist and with one physicist but is not so unusual among the social scientists. Note particularly the dominance of character in governing outcome,—and also the shift in social attitudes regarding the older woman when her pre-marital pregnancy (acceptable) became an illegitimacy (unacceptable). There are a great many points of interest in this story.

SUMMARY OF TAT INFERENCES

One of the first differences noticeable among these groups is the much greater length of the stories of the social scientists on the average than of the other groups. This is probably associated with their generally greater verbal productiveness. They also gave a full time range, including past and future, significantly more often than the others.

All of the groups are similar, however, in the proportion of responses which were more descriptive than narrative, and with regard to the types of outcome predicted, but there is a difference in the certainty of this prediction. The biologists are significantly less positive than either of the other groups. All of the groups tend to a combination of factual responses, with varying incursions of thinking and feeling, and relatively little use of make-believe or mythological stories.

These are formal aspects of the stories. One point about the content of the stories is common to almost all of them and this is that conflicts and difficulties are not often alleviated by chance intervention of any sort. There is occasional help from others, there are tragic stories of a death or illness but this is different. There is no reliance on "luck" or "fate" to solve problems, and in general it is implicit, if not explicit, that it is up to the person.

There are also major differences in content. Many of them show, in their stories, unease about intimate personal relations, but this'is of quite a different sort with the social scientists than it is with the biologists and physicists, who are strongly inclined to keep away from intense emotional situations as much as possible. With the social scientists what is shown is great concern with these relations.

Both biologists and physicists give indications of a considerable independence from their parents, and a feeling that this is right, whereas the social scientists show many dependent attitudes, much rebelliousness, and these are frequently accompanied by guilt feelings. Attitudes of helplessness are commoner among the social scientists. The biologists alone gave any indication of pervasive interest in their children or of particular meaningfulness to them of the paternal role.

None of these groups is particularly aggressive, so far as the TAT shows, but the social scientists are freest in any expression of aggression and the biologists the most restrained in this respect.

I had thought that the TAT should give some indications of the amount of drive for achievement that these men have shown, but this is rather strikingly lacking. It does occur, of course, as in some of the stories reported here, but there is relatively very, very little of it. This is important because it would seem that a major factor in their success has been the energy they have been able to put into their work. The TAT has been shown to discriminate in this respect with college students, in a study by McClelland and his associates, and this could be helpful in predicting success. It happens that I did not use all of the TAT cards which McClelland found most useful in this respect, but he and his associates were good enough to spend a day going over my data to check my impression on this point. They agreed that the sort of need for achievement which my subjects must have had, judging by what they have actually done, is not reflected in their TAT protocols, with a few exceptions. It is very possible that the fact that they *have* achieved has reduced the intensity of the need to do so. They do keep on working with great concentration, but this could certainly be partly habit, and a continuation of a form of living that has been and still is satisfying. Why should they change? This means, of course, that to learn whether need for achievement as shown on the TAT in young adulthood is related to eventual success, one must wait and see how men tested in young adulthood turn out. It is a job worth doing.

XIV

What Do They See in Inkblots?

THIS CHAPTER refers to the Rorschach Method of Personality Diagnosis, to give it its full name. Rorschach is the name of the Swiss psychiatrist who devised it. It is sometimes known as the "ink-blot test" because the test material consists of a set of ten standardized inkblots. These blots were made, as most of you have made them in school days, by shaking some drops of ink onto a sheet of paper and folding the sheet, so that a bilaterally symmetrical ink smear results. Some very fancy patterns can be made in this way. Of course, for the purposes of the test, the blots which Rorschach selected from an enormous number of them have been reproduced, so that the same ones are always used. Rorschach was not trying to devise a personality test, but was studying imagination. He was surprised to find that his inkblots could be used in making psychiatric diagnoses. He died soon after he discovered this, and the technique has been greatly expanded by others.

The Rorschach, like the TAT, is a projective technique and the remarks made about the difficulties in interpretation and in validation of projective techniques apply also to the Rorschach. It was the earliest of these techniques to become well-known and is still probably the one most extensively used. At first sight, though, it may seem even less probable that it can have any relation to personality than did story-telling. Again I cannot show you one of the ten blots actually used, but I have constructed some others. You see one in Figure 3. (I had so much fun doing this that it was hard to get back to work.) The instructions for the test are to look at each blot and to say what you see in it, what it looks like to you or what it makes you think of. Try it with the blot in Figure 3, before you go on to read what some other people have seen in it. You can write your answers on page 197 which has been left blank for you.

Figure 3

In the record of each person given below, I will number the responses for convenience.

When I look at this blot, I think

1. That the whole thing rather resembles a stylized design, with the two large lower sections a bit poodle-like.

2. Then, looking at just one of the "poodle heads" I think of Medusa and her snaky locks.

3. The tiny design in the center might be a butterfly.

4. Each of the top bits could be a sea-slug.

5. Or, with the blot turned on its side, the "poodle" section could be a person in a rather awkward pose with legs

apart and one arm stretched out and a sort of Fiji-islander type of headdress.

6. Or, turned right side up again, the 4 upper black parts, the 2 black dots and the white just around them could be the head of a ram.

My husband says:

1. That's the face of a monkey (the whole thing) in which the lower extensions are fuzzy jowls, the upper white spots are eyes, the black spots above are brow ridges, the black and white are the cheek pattern, the two black dots the nostrils, the little pattern below that the mouth, and the four black things extending down are whispers.

2. Now this also represents two animals which are curious zoological combinations of ducks and sheep and which are playing with each other. Each side of the pattern is one of the animals, at the top are their duck heads which are turned away from each other and which quite plainly have bills and foreheads and eyes; the rest of the large black spot represents the sheep bodies and you can see the two near legs and the big fuzzy tails on each one.

3. The two lower parts alone are each a poodle, with fuzzy head, legs and stubby tail.

4. This tiny bit in the middle between the poodles is a butterfly.

5. And it is also a cross-section of the spinal cord.

6. If you turn it upside down you can get a similar sequence of things but you also see something new. It's a mask this time rather than a monkey because it is less realistic, here are the eye openings (between poodle legs) and so on.

7. You also now have two Indian pots at bottom, each having a sort of lid with decoration on it there.

WRITE WHAT YOU SEE IN THE INKBLOT HERE

8. You also now have in this lower part of the figure two birds' nests and in each case a bird which can be seen either as just landing on the nest or as just taking off, depending on which you see as head or tails.

I showed the blot to Chico, our poodle, who is the only other person anywhere near. He was very haughty about the whole thing and a little insulted, but then we have never been able to decide whether he is a natural-born genius-type, or natural-born moron-type dog, although about 5:30 every morning when he wants company we strongly incline to the latter.

My publisher says:

1. First I saw a rather frightened grotesque creature with eyes too close together, clownish, amusing.

2. The two poodle dogs, very gay and abandoned and precipitate.

3. Then I saw two clutching hands hovering in a rather menacing fashion over that.

4. Silly, entertaining face.

5. The whole composite somehow oriental and a sophisticated compound of Chinese humor and French poodles.

6. Sidewise, defecations, gay ones.

7. Upside down, a weird, underwater seaweed.

Now look at Figure 4, and do the same thing, and then read what we see in it.

This time I see:

1. The whole thing as a Kachina Doll (one of the figures of gods that the Hopi Indians make for their children).

2. Then I notice the very black part just above the center and that looks like a bat to me, with wings outstretched.

Figure 4

3. The separate blobs on either side could be some uni-cellular animal.

4. The white streak down the center is a wand.

5. The upper extensions are two shrouded human figures, facing, with their arms stretched toward the center and sort of kimono-like sleeves on.

6. In the lower center the pale gray and white and the dots make an animal's head.

My husband says:

1. The whole thing is first a poodle head-on; the whole upper part is the head, the somewhat darker part is its rhinarium (nose to me), the white is the two eyes. His hair is parted, the spots off to each side are ear tufts and then the two front legs are here coming down together, a bit bow-legged and knockankled.

2. Second, it's a kind of goblin. The hobgoblin or dwarf has essentially the same features but he is bipedal and you don't supply legs and body in back. The whole body of the goblin is here, the face is the same and what were ear tufts of the poodle don't fit well with the goblin.

199

3. One also has here in the whole pattern a sort of flying butterfly although a rather clumsy one. The upper part is wings, the white patches are patterns on the wings, the side bits, if they fit in at all, are the front expansions on the wings which sometimes occur on butterflies and the things coming down are antennae, but they are much too heavy for the size of the butterfly.

4. Turned on its side I'm not satisfied with any form that I see. I could make a natural scene with a lake reflection out of it but it doesn't have the details to my mind for such a scene.

5. Looked at upside down the butterfly is better; it seems more natural with antennae upward rather than down although it is still too heavy.

6. Also now one has a figurine such as some of the Mexican or Peruvian ones, rather crude early pottery figurines. It is not realistic enough to be a live individual. There are the stubby legs and arms and abbreviated dwarf-like body and the relatively large head. Also characteristic of figurines, the features are poorly indicated but they are there with an elongate headdress on top. The two separate spots on the sides just don't fit in.

7. I also see this as something, possibly a hind-end of a squid breaking water. Squids do break water hind-end first and this upper part looks a good bit like the hind-end of a squid tapering down and with fins on the side and down here you can even see through the water traces of the tentacles. The black effects which were butterfly wings are the water thrown to each side by the emergence of the squid and the spots now become drops of water thrown up at the ends of the splash.

And my publisher says:

1. First I see a fine cross between a cat and a poodle, drawn up in its most alert and dignified and inquisitive manner, looking "down its nose" at me—self-important and yet sort of Alice in Wonderlandish, and conscious of being ridiculous like the black Queen or the Cheshire Cat.

2. I notice its paws drawn together in a rather effete, almost degenerate gesture.

3. The side blobs are sort of punctuation marks to the whole pomposity.

4. Sideways I can't seem to get anything except my other image on a horizontal.

5. Upsidedown. There is something sexy about this, the upper part is vagina-like, the other sort of building up to it.

Add to these what you yourself saw, and again you will note, as you did with the TAT, what a variety of responses there are to the same stimulus, and how many different things different people see. When you have read about the differences I found among different groups of scientists you will see that a number of things about my husband's responses are like those of other biologists (for example, the animal responses, of a type one wouldn't get generally, and the great care with form) and a number of things about mine are like those of other psychologists (particularly the responses with humans). I would not know about publishers and Chico refused to commit himself. (Perhaps he is the genius-type after all.) If few others see poodles in the blots it would be a fair inference that my husband and I are a bit preoccupied with poodles,—natural enough when we are likely to have our work interrupted at any moment by one. When Chico gets bored with himself he bounces his ball off our tables or puts it into a momentarily free hand for throwing for him, or just comes over while I type and lays his head on my lap,—usually with a long-drawn sigh which I am sure is meant to be heart-rending.

In giving the test, after the cards have been gone through once we go over them again. This part of the test is called the Inquiry and its purpose is to be sure that we know just where the subject saw each thing and how he saw it, and what it was about the blot that helped in forming the particular concept, for example whether the gray or black or white or colors made a difference, whether the shading mattered, and so on. It may be difficult to see things just the way the subject did, and sometimes, of course, we never quite do. In these groups if the response was of the technical sort that many biologists give I had to get someone to check it for me to see how well it fitted the area in which it was seen. For example, I had no idea what a "phyllocarid with abdomen upward and as seen from above" would look like.

As with the TAT, the record of the subject's responses is called the protocol. What do we do when we get all this down, and have seen things the way the subject does? First we go through a scoring procedure. Each answer is scored for the part of the blot in which it was seen, that is, whether the whole blot is used, whether an obvious large portion of it (like the "poodle" area in Figure 3), whether an obvious smaller portion (like the "sea slugs" in Figure 3), whether some unusual, less likely to be selected area or combination of them (like the "ram's head" in Figure 3 or the "bat" in Figure 4), or some of the white space (like the "wand" in Figure 4). Each answer is also scored for the striking aspects of it: the use of color or shading and its importance with regard to form;

the sharpness and clearness or the bizarreness and vagueness of the form; movement in the concept, animal, or human, or inanimate like a scarf blown by the wind; the use of space or distance in the concepts, such as landscapes or distant views. Then we note how many of each of these things there are in the whole protocol and various interrelations among them. Then, too, we classify and tabulate the content of the responses, how many humans, animals, plants, objects and so on are seen, and how many responses which many people see and how many are original. Of course we have a shorthand for this. (And of course there are several different "schools" of just how it should be done, but they do not differ very much.)

From all of these and their interrelations we can make a number of deductions about the person. It is easiest to illustrate this with the analysis of what parts of the blot the subject uses most. In general, people who tend to use the whole blot for a high proportion of their responses are people who like to consider a situation or problem in its total aspect, who like to generalize about things, the kind of person who sees the forest but not the individual trees. It is important, however, to note also how impressionistic the response is, or how clearly all the details of it are fitted together into a coherent whole, because this also has meaning. Is it a sweeping judgment type of response, or a careful generalization on the basis of all of the facts? People who tend to give large numbers of responses to the large and obvious details are likely to be of the very practical sort, who take care of the obvious things first, and are good at the ordinary details of living,— people who see the trees in the forest (perhaps to pick out the ones that would make good beams) but do not pay much attention to the shrubs or delicate flowers and are not particularly concerned with what type of forest it is. People who spend a lot of attention on the small details, who look at unusual aspects of the thing may be people with original minds or they may be pedantic and sticklers for detail; other aspects will clarify this. Finally, people who use lots of the white spaces seem to be people who like to do things in their own way, which is generally not the usual way. But all of these interpretations are subject to qualifications on the basis of other aspects of the responses. In interpreting the Rorschach, you can never consider any one aspect separately.

From analysis of the relative proportions of these locations, the sequence in which they are used, and the relative complication and excellence (in terms of how well the concept fits the blot) of the different types, one can usually tell a good deal about the way the subject goes about his work. I did not make any rigid experimental test of this. (It would not be an impossibly difficult thing to do. The Rorschach end would be easy enough, but a major problem would be the criterion for how the work was customarily done.) I can give some examples, however, of the sort of thing I mean. Subjects with a low percentage of good whole responses are likely to be subjects whose major contributions are in the way of piling up evidence upon evidence on particular aspects of their fields rather than subjects whose major contributions are in broad theory and generalization. (I have the strong impression, although not enough data to assert it, that a good part of the controversies over classification of animals and plants results from differences of this sort in the persons doing the classifying,—that is, given the same array of specimens, people who give many whole responses on the Rorschach are likely to put these into fewer different species than people who give many detail responses on the Rorschach.) Other relations can be shown. I suggested to one man, for example, that while he was meticulous and critical about detailed parts of his work, he was rather given to making sweeping generalizations, sort of off the cuff, and that they wouldn't be as good as the rest of his work, although on second thought he might produce good generalizations. His reply was, "Oh, boy, do I! I have finally learned never to publish my first conclusions." The subject with the largest number of white space responses, many of them originals, that is, responses that not more than one person in a hundred would think up, is a physicist who says that he doesn't like to do anything unless he can do something completely different. "I have that ability to think up unusual things in odd fields." Another physicist remarked of him, "He can provide ideas enough for 100 people. That's what they have him there for."

The relative use of color and movement tells us something about the subject's sensitivity to outside stimuli and to internal preoccupations; his use of shading something about his sensitivity, anxiety, and mood swings. Subjects who give many replies, based only on the form of the blot or portions of it, and disregarding other aspects such as color or shading or using them only secondarily, are people with considerable reliance upon intellectual control. We also record whether or not the forms are good ones, that is, congruent with the blots, and this tells us a good deal about the effectiveness and the tightness of the control. Like most other things there is an optimum amount of control. Too much or too little indicate difficulties but of different sorts.

There are still other things to consider,—what responses come first on each card, how long a pause precedes the first response to each blot, how many responses there are to different cards, and altogether, whether there are differences from one card to another and so on. Furthermore, after all of these details are looked into and considered in relation to each other, one then goes through the responses again, one by one, and tries to analyze them qualitatively. All of this may take a long time. One cannot do very much with very short protocols, and sometimes very long ones are hard to work out just because of the superabundance of the material. I have taken as much as a day on a single record and I usually expect to spend several hours on any record if extremely careful analysis is important as it was for all of these subjects. There are shortcuts, but as is true of most things, the more time you spend on it the more you get out of it.

In comparing one person with another or one group of persons with another, one can of course take up all of the types of scoring mentioned and compare them directly. (We have no statistical technique as yet for overall comparison and that still has to rest on descriptive summaries.) There is, however, one major difficulty here, and that is that almost all of the different scoring categories are greatly affected by the total number of responses the subject gives, but the relationship is never a simple one. We can't take care of this problem by just putting everything in percentages either, for various reasons. For example consider the number of whole responses, or responses in which the whole blot was used. Obviously a few of these are easy to give, but it is extremely difficult to work out a great number of them. Suppose one subject gives 5 whole responses out of a total of 15 responses. This would give him

33% of whole responses. But suppose another subject has given 75 responses altogether, and again 33% of these are whole responses. That means he has given 25 whole responses, an extraordinarily difficult feat which the percentage figure does not suggest at all. Several proposals have been made for getting around this but I have used the system developed by Munroe which makes it possible to compare a scoring of one protocol with that of another, with the scoring of each adjusted for total number of responses before comparison.

It is fortunate that there is some means of handling this problem, for one of the ways in which these groups differ greatly is in the total number of responses they give on the average, to all 10 of the cards. (I should point out that most of the inkblots actually used in the Rorschach are more complicated than the ones I have constructed as illustrations, and it is easier to give more responses to the Rorschach blots. But people vary enormously in this respect anyway.) For the biologists, the average number of responses is 22, for the physical scientists 34 and for the social scientists 67! Among the physicists, the theoretical group gave many more responses on the average than the experimentalists. We think of such persons as being freely productive, but this is associated in the theoretical physicists and the social scientists, with a fairly marked carelessness about any system in the handling of new material and some general uncriticalness. The biologists, on the other hand, tend to be very careful; they are extremely concerned with the form qualities of the blot, and critical of their own responses, and on a number of aspects they show a much stronger reliance on intellectual rational control than do any of the other groups. They are, on the whole, probably the best integrated and the least prima donnaish of the lot, although there are prima donnas among them.

The physicists seem, as a group, to have a good deal more anxiety than the biologists, that is, they are much more easily disturbed without obvious reason, and they do not have very good techniques for handling such disturbances. The social scientists show no general consistency in this.

Both biologists and physicists are less interested in other people than the social scientists are and this is particularly true of the biologists, but they seem to be able to cope with other people rather better than many of the physicists. The social scientists are extremely sensitive to other people, and much more concerned personally with such matters as dominance and submission and the need for support and affection than the other groups are.

The biologists show very little in the way of aggressive tendencies although I should judge them to be a very stubborn group on the whole; the physicists are somewhat more aggressive but not markedly so; the social scientists, particularly the anthropologists, seem to be quite freely aggressive.

Differences in the kinds of things seen in the blots are quite interesting, too. The biologists saw a lot of animals and animal structures, more than the other groups. This is not surprising for the blots lend themselves readily to such interpretations, and many people do give a number of animal responses. The physicists saw more objects, and were much more likely to see inanimate things in motion, or to see things in three-dimensional space settings than the other groups; the social scientists saw many more people than the others did and they were more likely to mention clothes and food than the others.

All of these differences make quite a lot of sense in view

of the things with which these men concern themselves all day long. It would seem that they have ways of looking at things which apply not only to what they do professionally but to such a different process as looking at inkblots. On the other hand, all of the groups have a few things in common. They are somewhat more likely than most groups to generalize, to see things in a larger framework, but this is pretty varied among them. What is even more characteristic, however, is a tendency to look at unusual details and this is perhaps their outstanding difference from the general run of people.

Below I have listed the responses to the same Rorschach card from a botanist, a biochemist, a theoretical physicist, an experimental physicist, a psychologist and an anthropologist. I have tried to select these from protocols that are most typical for the groups, but of course no one individual is just typical.

Response of A, a botanist:

1. The whole doesn't remind me of anything. But the pygidium reminds me of a horseshoe crab. There is certainly nothing like that on limulus. It looks like a trilobite tail or horseshoe crab tail; it would have to be a trilobite, not limulus.

This is the kind of response which requires consultation with an expert in order to judge its adequacy. I never heard of a pygidium before. It is fairly typical of the biologists.

Responses of B, a biochemist:

1. Two dancing bears. The trouble with that one is it looks too much like the bears and you can't think of anything else. Bears with injured feet and red caps.

2. There is a white arrow head in the middle.

A number of people have the kind of trouble he suggests in his first response. If one thing is seen with special vividness, it is hard to see anything else. He manages here, by looking at a small detail, instead of at the whole thing.

Responses of C, a theoretical physicist:

1. Two people talking or playing patty-cake, I don't know which. It would have to be people from Mars because the shape of the head is very peculiar.

2. One always gets phallic symbols out of these things of course. This is a double one, both male and female.

3. I can't quite figure what kind of a 4-footed animal this is but the head is over here of course. Probably a furry black dog.

4. And there's a butterfly here.

5. The center would be the main works of a turbine with the bearings here and here. It definitely has to be in motion.

6. Are these printed or made? I wondered about the mottling, and the circular symmetry. They tried hard to get it, it must mean something. I was trying to figure if they were arcs of parallel or concentric circles.

The much larger number of responses to each card is typical of men who think verbally as this man does. Note the number of different classes of things he refers to, people, symbols, animals, and finally a turbine before he begins to wonder about how the things were made. The final remark, the comment on "arcs of parallel or concentric circles," is of a sort I have had only from physicists. The wide range of content, the ease of the shift from one to another, and the changes from use of whole card to different details, but in no particular order, are very characteristic of this group.

Responses of D, an experimental physicist:

1. This looks like 2 moles that somebody mashed. You know, a mole that goes underground. An animal that's been mashed and that's blood. It has a tendency to look furry because of these colorings in stripes.

2. You can imagine different things. These are animals of different forms like a bear or something like that. If you cover half, you get a very definite imprint of a cross section of a bear at the top.

You can see how differently the experimentalist responds. The "mashed moles" is an unusual response, which has a very unpleasant feel to it. His second response is a better one, but even there he has a little trouble, and has to cut off the bottom of the card, although most people can use the whole card in the bear response.

Responses of E, a psychologist:

1. There are two clowns playing patty-cake on the stage. Perhaps a ballet performance. They have red caps and blouses, under some kind of caracul costume, a caracul coat. It's caracul because of the light and dark reflections you get and it's black.

2. There is a smear of blood as if a big bug had been squashed down at the bottom.

3. Up in between where the clown's hands are there is a kind of an inverted cone, it looks like what's left of a huge lily after the petals have fallen off.

4. Over on one side there is a terrier. I guess he is sitting up kind of erect, sitting up on his rump with his forelegs up and staring out rather proudly, interestedly. He's one of these dark brown terriers with sort of reddish spots. He's quite a

proud and happy fellow.

5. Down at the bottom there is something which looks more or less like a vaginal canal or surgical incision in the abdomen or something of that sort.

Like the theoretical physicist, the psychologist starts with a human response and includes, in different order, animal and symbolic responses. But he sees as a cone what the physicist saw as a phallic symbol; on the other hand he gives a female sex response. Seeing either a vaginal canal or surgical incision is quite suggestive as to his attitudes towards female sexuality, wouldn't you think? Responses 1 and 4, though, are particularly "good," well elaborated, with pleasant overtones.

Responses of F, an anthropologist:

1. This suggests two clowns or two Shamans. It gives me a Siberian feeling in some way, they are striking hands in some sort of dance. It has a Russian effect to it, with red caps and red feet, sort of bloody feet. The coats suggest bear coats.

2. The red gives me the blood clot idea but it doesn't fit into any scene.

3. The head of a man with the face going out between the two red blots, with strange shoulders and arms.

4. Also there are tied arrow heads out of that middle thing between the hands of the dancers. Sometimes that part is the hands tied together and then it is tied arrow heads that they may be holding.

5. A skull like a rabbit skull out of the white between the dancers, going into the red. This might be the nose of the animal, possibly with dried blood on it. It's a very poor skull.

6. I can see in the bodies themselves, these black bodies, the head of an animal, a bear on each side, with one red eye

showing and the nostrils indicated. One looks like a suckling cub, the other not so much. The heads go down into nothing in particular or else they are foreshortened bodies on the cubs which are not foreshortened on the larger figure.

7. The white thing is a little bit like a flat fish of some kind or a sting ray.

8. The lower red, and the white between the red and black figures has a sort of anal or bloody anal or somewhat surgical suggestion or it's vaginal.

Again we have the large numbers of responses, and the likenesses to the replies of the psychologist are quite striking. The differences, though, are illuminating. The mention of Shamans, or medicine men, is a technical touch, of course. Blood is the second response in both cases, but then the anthropologist sees the head of a man. What the psychologist sees as a cone is now arrow-heads, or it may be that the hands of the dancers are tied together. This is a significant restriction of human activity. Then he, too, winds up with a vaginal or surgical response. Note, however, that this series of responses is unlightened by such a pleasant one as the proud terrier. This man is more immersed in his problems, more tied by them.

XV

The Subsidiary Study

IT WAS WHILE working on this part of the study that I ran into
the perfect exemplification of "Murphy's law" at one university, where everything that could go wrong did! It was all very
embarrassing but it was fortunate that it happened to me and
not to one of my assistants. I will tell you about this later.

I mentioned this part of the study in Chapter II when I
started a discussion of sampling problems. The reason for
doing this subsidiary study was that I had no check on how
closely eminent men resembled other men in the same fields,
nor are there any background data for high level workers in
science generally. Since I could only study a few men individually, my samples would be very small for comparisons.
The two parts of the study were carried on simultaneously,
of course, but it is simpler to discuss them separately.

What I wanted for this part of the study was as large a
group as possible of successful workers in the biological, physical and social sciences. I wanted them also to be as diversified
as possible. Since I wanted them of already demonstrated
competence in their own fields, I did not want students or
even graduate students, the usual happy hunting ground of

psychologists. I wanted people who were holding down good jobs, and I wanted them to be reasonably representative of the field. All were to be given the Rorschach as a group test.

I needed information about the location and distribution of scientists in different fields in order to plan the sampling procedures. The Office of Scientific Personnel, established by the National Research Council, has made surveys of the country's resources of trained scientists. They were very helpful and cooperative and supplied figures which indicated that about 80% of active research scientists in the country are connected with academic or research institutions, the others being chiefly in industry or government. Since all but one of my individual subjects were connected with universities or research institutions, it was in these that I decided to look for subjects for the group study.

In this, too, Dr. Trytten of the Office of Scientific Personnel, proved most helpful, as he was able to give me figures, collected over a 12-year period, for the number of doctoral degrees in the different fields given by all the universities in the country. This was helpful because it seemed evident that the institutions giving the largest number of degrees in any field would have the most specialized and probably the largest faculties in the field. These, then, would be the faculties from which I would try to get my subjects.

There were also other things to be considered. No one had made any studies of this sort of university faculties, so I had no idea whether or not the nature of the university, i.e. public or private, or it's geographical location would have any effect on selection of the faculty so far as personality characteristics go. These conceivably could make a difference. I decided therefore, the first year, to take the public and private university in each of four geographical areas, which had given the largest number of degrees in biology during the period for which the data were available. I followed much the same system during the second and third years for the other groups studied, except that I did not attempt to include any southern universities, since few of them give many doctorates. (I did not get farther south than Maryland, even, the first year.)

For this part of the study I did not exclude any subjects on the basis of age, sex or national origin. There are a number in the group who are as eminent as the men studied individually, but who were not considered for that because of age or national origin. There are also two psychologists who were the ones eliminated from the group for individual study when I tossed a coin to see which one of three to take.

In all, there were 188 in the biological sciences (including 18 women), 65 in the physical sciences (including only 1 woman) and 129 in the social sciences (including 16 women). The universities from which these groups were obtained are indicated in Table 13. The proportion of women in each group is about what it is in the general field.

When the Rorschach is given as a group test, slides are substituted for the usual cards, and these are thrown upon a screen in the usual manner. I used Kodachrome slides, as they are the simplest to carry around. The lighting is adjusted so that the subjects can see well enough to write in the standard booklets which have a separate double page for each of the 10 slides. Each slide is exposed for 3 minutes and the subject writes down, on the right hand side of the page, what he sees in it. Then the slides are shown for a second time. On the left page of each set is a reproduction of one of the blots. This time each slide is left on for as long as is needed for the

Table 13

University Faculties Participating in the Group Study

UNIVERSITY	BIOLOGICAL	PHYSICAL	SOCIAL
California (Berkeley)	X	X	X
Chicago	X	X	X
Columbia	X	X	X
Cornell	X	X	
Harvard			X
Iowa			X
Johns Hopkins	X		
Maryland	X		
Michigan			X
Ohio			X
Stanford	X		
Wisconsin	X		
California Institute of Technology		X	
Massachusetts Institute of Technology		X	

subjects to mark on the reproduction of the blot, the area in which they saw each of the responses, and to note what aspect of the blot, form, color, shading, movement, contributed most to their response. Naturally the results when the test is given this way are not exactly comparable to results when the test is administered individually but in my opinion they are near enough to use cautiously in comparisons. (I say in my opinion, because this has not yet been adequately checked experimentally. Such evidence as is presently available tends to support the comparability of the two forms of administration and my own experience is also strongly in support of this.)

I did the testing at three institutions for the biologists, at two for the physicists, at six for the psychologists and at all for the anthropologists. Testing at the other institutions was done by various of my colleagues. The testing itself, is a simple matter, but rounding up subjects is not. The technique for this differed somewhat from place to place. When I was going to be on the campus of any university to see one of my individual subjects, I usually postponed having the group testing started until after I had been there. During my visit I then went around and saw the chairmen of all the departments involved. In biology, for example, this meant seeing people in the departments of anatomy, physiology, zoology, botany, genetics and any others. I explained the project, told them who was to do the testing, and tried to work up some interest in it. Then one of the local people took over, and did the actual work of finding out who would come, when would be the most convenient time, and where it could be done. These are all very time-consuming details and this is one reason why, whenever possible, I had someone else do the work. I also felt that a local faculty member might be able to get better cooperation from his own colleagues than I could.

Once the test was given the records were sent directly to me. They were then scored by my assistant,—I had two different ones during the course of the project,—and eventually I went over each record also. When my assistant and I differed on scoring that part was checked with great care. We differed rather little, as a matter of fact, so that I did not have any recalculations to do. Having two scorers reduced the amount of subjectivity in the scoring which is large in this test anyway. It is also the case with me that I cannot work happily with data which I do not know intimately at first hand, and there are many aspects to the Rorschach which do not appear in the tabulations of scores for each person.

I will never forget the last batch of scoring I did,—that for the social scientists. I had done some before leaving New York for the summer, chiefly the records on which my assistant had had particular trouble, so that we could discuss the difficulties in person rather than by mail. The rest I had put aside to do later. Soon after going West I decided to go along with a fossil-hunting party on a trek to Wyoming, thinking I could spend my time scoring Rorschachs while the rest were looking for fossils,—a very hot job. We expected to spend most of our nights in towns, but took along a minimum of equipment against the chance of camping out for a night or two. We had an umbrella tent for my husband and me, a pup tent for the boy we borrow summers (since we have no sons and a New Mexico place needs a boy on it) and I believe the other three, a professor from the University of Wyoming, a student from California and my husband's assistant, had one pup tent among them. We also had a camp stove, one chair, and our sleeping bags.

As it turned out we spent two weeks camped in one spot, on a treeless, waterless sagebrush plain in western Wyoming. It was wonderful. The rest were gone all day, at first, and I had no interruptions and turned out mountains of work. I sat in the front seat of the car, the tent being much too hot during the day, and scored Rorschachs all day long until I had ink blots before the eyes. But it was easy to rest your eyes in all that space and there were almost always antelope around to watch when a distraction was needed. They did not come close to camp but they are very curious creatures and occasionally would slip up to the men in the field to see what was going on. It can be quite disconcerting, working on hands and knees on a badlands slope, to have an antelope start snorting in your ear without warning. Water was about 20 miles away but we were all old hands at making dry camp. That was something of an experience, especially as four of the party came down, one after the other, with what we dubbed the "misery," because there never was such an aching. There was no doctor for a hundred miles or more so we sat it out, especially as the symptoms were more miserable than alarming. After three days we did decide I had better take the boy to a doctor (he was the first to come down) but he demanded 6 pancakes for breakfast the next morning so we decided he was on the road to recovery. Then two of the men came down with it one after another. We found out that our "misery" was a particularly virulent flu when I developed a temperature of 104 and was hastily driven to a doctor. Fortunately that was the day we were planning to leave anyway.

But back to the testing. In addition to asking each subject to note on the front of the record booklet (or on a special sheet prepared for this for the social scientists) age, sex and field of specialization, I also asked them to state whether they preferred research or teaching. The social scientists were also asked to state how much they knew about the Rorschach so I could check the effect of knowledge of the test on the nature of the responses. This group was asked not to put their names on the tests (since my assistant and I would know a number of them and it seemed better not to know at the time whose test we were scoring) and to do as they liked about putting them on the other sheets. The others were told to put their names down or not as they wished. Most of them did put them down.

I should, I think, make it clear that I realize that it is quite a lot to ask a busy faculty member to take the time and go to the trouble of going to a particular room, probably in another building, at a given time, in order to be subjected to a test of which they had probably never heard, and the purpose of which was obscure. And this was usually at the request of someone they had never heard of. That so many were willing to do this is an indication of the genuineness of their interest in science,—and of the persuasive qualities of some of their and my colleagues.

It was more difficult to get subjects among the physicists than among the others, for this as well as for the main study. Again, they are a busier lot, but the extent of the difference and the nature of some of the things that happened supports the interpretation that this greater reluctance in part reflects the greater amount of free anxiety that the test results themselves show. It is quite possible, too, that the general harassing to which so many of them have been subjected played more of a part here than in the individual studies. At one university where I gave the test, for example, the department chairman, although personally interested, declined to assign a room for the test at first on the ground that I would not be able to get any of his faculty to come. I talked to most of them individually, and got a number to agree, however, whereupon the chairman willingly made the necessary arrangements, came himself to the test, and when he left was still surprised that so many had turned up! In another university, after all arrangements had been made, and there had been a number of personal interviews, the local psychologist who was doing the testing sent around a mimeographed final reminder of time and place. Unlike previous notices of the sort, it was not signed. The chairman in physics promptly told all his department that there must be something wrong because this notice was not signed and not to go. A number of them did go, fortunately, and also fortunately the chairmen in physical chemistry and geophysics had not had the same reaction.

When I first broached the subject to the chairmen of the various departments in psychology, there seemed to be a general feeling that it would be much better for me to give the test myself. At one university no one seemed to care who did the testing so I left it to the psychologist who had done the work there for me the two previous years. At most of the universities where I tested psychologists and anthropologists there was usually some informal discussion of the whole project after the test, and of some of the findings that were already beginning to be worked up. This, of course, was lots of fun for me, but there was one experience which still gives me the cold shudders.

This was at a midwestern university. I had already written to a clinical psychologist there whom I knew and he had taken the matter up with the people in the department. He was not sure how much cooperation I would get but thought it would be all right. The chairman of this department, a man I had not met before, is a particularly rigorous, and very good experimentalist. (He was one of those eliminated from consideration for the individual group by the coin tossing.) I was reasonably certain that he would personally disapprove the Rorschach because of the lack of checks for validity and reliability, but he was extremely courteous and helpful as were all I saw there and arrangements about time and room were easily made. The room selected was one that had been partitioned off from a larger one. The other half would also be in use at that time for a class. No one thought that this would matter, although the partitioning was not entirely sound-proof. There was already a projector set up in the room, but it had to be dismantled and one that would take Kodachrome slides put in. We were about to start, when I ran through my slides to be sure they were in order and found, to my horror, that my last slide was missing. Presumably I had left it in the projector at the university (some hundreds of miles away) where I had last given the test. I had no extra set with me, nor could any be found at the university or at any clinic in town. What to do? I had to have all 10 cards of course. They had, fortunately, the kind of gadget that can pro-ject printed matter so one of the clinicians went and got his set of the standard cards while we changed projectors again. Then the question was should I use my 9 slides as usual and then that projector for the 10th? I decided not, since there was some difference in the size of the images by the two systems and I thought it best to keep them all the same for one test. In either case this test would not be exactly the same as the others. Finally we were ready to start,—we thought. I adjusted the floor lights that had been brought in, so there would be enough light to write by but the projected image would not be affected, and turned off the ceiling lights. There was a howl from the next room. It seemed that when the room had been partitioned there had been no partitioning of the lighting system and the class next door needed light on that gloomy winter afternoon. Hurried consultations! The professor, bless him (I never learned his name), upon learn-ing that there was a vacant room on the floor above which we hadn't used because it couldn't be darkened, most equably took his class off there, and the test finally got under way. Next day, when I had another but very small group, I did something still different. We just sat around a table and I used the cards, holding them upright. When I went over the results I paid very particular attention to records from this university but I could not find any way in which they seemed to differ from the others which could possibly be laid to the vagaries of the test situation (which I strongly deplore none-the-less). The only possible explanation for this is that the subjects were extraordinarily patient and cooperated wholeheartedly in spite of the difficulties, and that this saved the situation. Certainly this was Murphy's law in operation,—but I will never be the same again.

Before discussing differences between the subjects in different scientific fields, it may be well to discuss first other possible differences. I checked for sex differences among both biologists and social scientists and found none, so have reported the sexes together.

Regional differences I checked only for the biologists. There were some, they were very scattered, and they did not fit any sort of a pattern. In view of the fact that the two universities in each region were not selected because they were typical for that region but because they had had the largest number of doctoral students, it seemed unwise to try to decipher any possible regional differences in this way. What the comparisons did show was that there might well be regional differences so that future samples also should include subjects drawn from different parts of the country.

There were some striking differences between different universities. The most noticeable is a difference in the concentration of individualists on the staffs. Some universities seem to have them in all departments tested, some in only some and some not at all. Apparently there is a real difference in the amount of tolerance which universities or university departments have for prima donnas. This is a rather interesting point. Extremely able men are often rather independent in many ways (and sometimes eccentric) and so concentrated on their own interests that it is easy to see that this could introduce administrative difficulties. At the same time this very individualism is what makes for scientific advance, and from a practical standpoint attracts students to the university. On the whole the public universities seem to have less of this sort of tolerance than the private ones, although there are notable exceptions. It is not too surprising that state universities which are dependent upon usually conservative state legislatures for their funds may be hampered in this respect. Yet here lies a danger which we should be very conscious of,— pressures towards conformity of any sort always militate against original work.

The University of California has been a marked exception to the tendency on the part of many state universities to conservatism so extreme as to result in intellectual stodginess. Its reputation for having a brilliant and individualistic faculty, which has made great contributions in all of these fields is well-deserved, and my own results with the Rorschach show clearly that it has harbored an unusual number of particularly fine and extraordinary minds. The recent battle with a group of reactionary regents (I should like to have Rorschachs on them, but I think I can guess what they would be like) over the issue of academic freedom has had and doubtless will continue to have disastrous effects for the university. (Other universities have profited by getting former University of California faculty members for their own staffs.) It is far from true that independent thinkers are anti-social, they are much more likely to be stable members of society in ways that are important. Social stability which can also permit valuable advance is not characteristic of sheep-like groups.

I did not find any differences ascribable to differences in rank in the university (instructor, professor, etc.) but I checked this only in the biologists where the group was large enough to subdivide in this way. The effect of age is practically negligible for the total group. About the only item in which it may be important is a slight decrease with increasing age in the number of responses involving human movement.

I made some attempt to find out whether there seemed to be any differences related to preference for research or for teaching, but I found none. Many of these subjects like best to do some of both, but there are some who teach in order to have a place in which to do research, and there are some who do research as a means for advancement in teaching position. The attitudes of university administrations differ on this. In many universities of standing, advancement in position in the university is strongly affected by number of publications, or other research output. This can have the advantage of making it easier for men who want to do research to do it. It also has the disadvantage that it may result in the production of lots of unimportant research to the end of getting more papers in press, so the titles can be shown the dean, and the further disadvantage that there are some first-rate teachers who are not interested in research, and whose time would better be spent in teaching. Other university administrations discourage research on the part of the faculty, because of the time it takes and because research facilities cost money. These universities, however, usually do not get large numbers of graduate students or the pick of them.

There are some differences among different subdivisions of the main groups but these are very varied and also less than the differences between the main groups.

Special mention should be made of the problem of knowledge of the Rorschach. There were 62 psychologists who had no knowledge of the test (other than casual reading), 22 with some knowledge of scoring and 20 who are experts in the use of it. These groups were compared, but there is a further difficulty here. That is that, of course, it is the clinicians who know the test, so that it is almost impossible to separate the differences which result from being clinicians rather than some other kind of psychologist from the differences which result from being expert in the test. Most psychologists would agree, I think, that there may well be personality differences between psychologists who go into the clinical field and those who do not. In any case, clinicians differ from non-clinicians, and experts differ from non-experts in the same ways on the Rorschach and these are chiefly in the greater use by clinicians and experts of human movement responses, in care with form accuracy and in generally more balanced performance. Nevertheless all kinds of psychologists are more like each other than they are like other groups and the anthropologists are sufficiently like the psychologists that they can be grouped with them for this analysis.

What, then, are the major differences between the scientists in these fields as they show up on the Group Rorschach? It was extremely encouraging to find that they are practically the same differences as those that were found on tests of individual scientists which are reported in the last chapter.

Again the social scientists are way ahead of the others in total number of responses. (This had an economic aspect for me. I had been paying my assistants so much per record for the scoring, that being simpler for them than keeping track of the number of hours they spent on it. But when these records came in, it took so much longer to score them that I had to pay more per record.)

Again the biologists show up as very concerned with form on the blots; the physical scientists as being much more concerned with 3-dimensional space and inanimate motion than the others and the social scientists as extremely concerned with (in fact, practically haunted by) people.

The major differences in content are the greater use by biologists of anatomical and plant responses, the greater use by the physicists of responses involving explosions and snow and ice; and the greater use by the social scientists of responses involving humans, or human details, or clothing.

On the Group Rorschach a general estimate of overall adjustment tended to show that the social scientists and particularly the psychologists were better adjusted than the other groups. (In the individual studies, the biologists had come out much the best and the social scientists the poorest.) It may well be that the fact that the Rorschach and our concepts of maturity have been developed by psychologists has something to do with this. Naturally we would be inclined to make a general ideal of what seemed to be ideal for us.

The subsidiary study, then, demonstrated that differences between men in different fields will be shown on the Rorschach, whether the men are selected from the best in the field, or from those who are competent in the field, and whether the test is administered individually or as a group test.

What about the differences between the best in any one field and others in the field? These varied in the different groups. Among the physicists I found no differences of importance. Among the other two the men individually studied tended to have extremer degrees of the factors that particularly characterized the men in the field. The eminent biologists were even more concerned with form, more controlled and generally better balanced than the others in the field. The eminent social scientists show almost an opposite picture. They are more productive than the other social scientists, and more original, but they are also less controlled and give more evidence of severe personal problems in adjustment.

XVI

What Does It Mean for You?

LET US TRY, now, to pull all these observations together, and see if they make a meaningful pattern and what their implications are.

One thing seems clear. Scientists are people, not rational automatons. They differ from other people in terms of what they do, in the things that give them satisfaction, more than in terms of completely special capacities. There is nothing you can say about them as persons that you cannot also say about some people who are not scientists. And there is almost nothing you can say about a man in some particular field of science that you cannot also say about someone in another field of science. In spite of this, there are patterns, patterns in their life histories, patterns of intellectual abilities, patterns of personality structure, which are more characteristic of scientists than they are of people at large, and some which are more characteristic of special kinds of scientists than of other scientists.

There are distinct patterns in these life histories. These scientists come from rather selected families, since half of them had fathers who were professional men, and none of them had fathers who were unskilled laborers. And none of them came from Catholic homes. This pattern is about the same in all of the groups, except that there is a higher proportion of professional fathers among the theoretical physicists, and a somewhat lower proportion among the experimental physicists.

Not all the sons of professional fathers become scientists, but this background more than others, seems to have a predisposing effect. Why? By and large the intelligence of people in professional occupations is relatively high, and so their children would be expected to be relatively bright, hence there may be some hereditary factor. But the inheritance of intelligence is a very complicated matter, and not too well understood, and I think here of relatively less importance than other things. What seems to be important in the home background is the knowledge of learning, and the value placed on it *for its own sake*, in terms of the enrichment of life, and not just for economic and social rewards. This high evaluation placed on learning and on intellectual satisfactions was also operative in many of the homes in which the father was not a professional man. The few scientists whose homes lacked this had always had close contact with some one else, usually a teacher, who held this attitude.

More than is usual, these men were placed on their own resources. It happened in different ways, in different cases. Some lost a parent early in life; some had serious physical problems; many were eldest sons, although the bearing of this is not clear. Even as children, though, most of them had intense private interests, and except among the social scientists, these were usually shared by only a few friends. Most of them were inveterate readers, and most of them enjoyed school and studying.

Very early there were differences in the things that interested them. Many of those who became physical scientists and almost none of the others were early involved in gadgeteering of one sort or another. Many of those who became biologists, and a few of the others, particularly anthropologists, were extremely interested in natural history from early childhood. Many of those who became social scientists and very few of the others went through a stage of planning a literary life. A few social scientists became so primarily in an attempt to work out a desire to be of service to others, but I do not find any evidence of such a motivation in any of the other scientists, and it is relatively rare even among the psychologists in this group. (But these are mostly experimentalists; it is a commoner motivation among clinicians.)

This leads me to comment upon the opinion, held by many people, that the scientist is a completely altruistic being, devoting himself selflessly to the pursuit of truth, solely in order to contribute to the welfare of humanity. I do not intend it as a derogation of men whom I cherish when I say that this is, in my experience, not really the basic motivation for any of them, and as additional motivation it is more often absent than present. That they do, in fact, expend themselves in activities which are a very real contribution to humanity is the good luck of society. That what they do does make such a social contribution may, indeed, give great satisfaction to them and may even make the meagerness of their financial rewards more tolerable. (I do feel that this meagerness does not speak well for society which derives such great profit for so little expenditure of support or recognition.) There are those among them, however, who have never given spontaneous thought to such considerations. Indeed, as a psychologist, I must say that I should be very doubtful of the emotional health of any individual who thought he "gave up" all personal interests to "serve humanity." As an additional motive that serves to channel personal interests in a particular way, as was the case with one of the psychologists whose life you read, it is a different matter. This is genuine dedication and a healthy one.

On the other hand, that scientists are now being forced to consider the social repercussions of their work is an excellent thing, both for them and for society. The man so completely immersed in his work as to ignore social problems, however important his work may be in the long run, is less than a man and so has failed the thing most requisite upon him.

Level of intellectual functioning is very high in this whole group, and there are pattern differences here, too. Social scientists and theoretical physicists tend to relatively higher verbal than non-verbal abilities; experimental physicists tend to relatively lower verbal abilities than non-verbal, and the anthropologists tend to relatively low mathematical ability. It is very probable that these abilities were a factor in choice, not only of science, but of the particular science. That they would not be in any case the decisive factor is shown by the fact that there are exceptions to all of the generalizations stated above.

In general personality structure, it appeared that biologists have an orientation which strongly emphasizes reliance upon rational controls. Both of the other groups tend to be uncritical people, with much less insistence upon rational control and rather less of it. The physicists are often anxious and neither interested in people nor very good at relating to them in general. The social scientists are deeply concerned about

233

human relations, and also troubled by them in a way quite foreign to the other two groups, who prefer to maintain some distance, and generally succeed in doing so quite guiltlessly.

Though these generalizations are valid within limits they do not tell us exactly why these men became scientists, and not something else. Special abilities require special activities for their satisfaction, but the same type of activity can be found in many occupations. Choice of science is dependent upon other matters, primarily upon needs of a more general sort that are best satisfied by this profession. (That it is satisfying is evidenced by the fact that none of these men had any wish to change professions.) The occurrence of the situations noted above gives us some hint of the kinds of needs that are involved.

One of the first things one notes about scientists is the fact that a large part of their time is spent in thinking about things, in a question-answering way. They want to find out something, and all of their activities are designed to bring them answers to questions. (Of course a good part of the trick to being a first-rate scientist is in asking the right questions, or asking them in ways that make it possible to find answers.) One way of putting this is to say that they are curious, even though each may be chiefly curious about very specialized things and not very curious about all the other things in the world. All children are curious, I think, but not all adults are. For one reason or another, many adults are unable to be curious to any great degree, or are able to be curious about few things. It is true, of scientists, too, that some of them seem to be curious only about some things, but most of them have a more general sort of curiosity. This limitation of a healthy and intelligent person's normal reaction of interest in the world around him may result from repressive training, from discouragement of questioning by weary parents or teachers, from an adult attitude of know-it-all, and insistence on a child's conformity. It may also result, of course, from special emotional problems. An intense channeling of curiosity into a narrow field can also be the result of neurotic problems. There are individual differences in this sort of reaction to start with, but we know little about these. That some children get major satisfactions from physical or social activities involving little questioning or thinking is clear. But there are others who do not find enough in physical or social satisfactions. The effective functioning of any bodily system is satisfying and this applies to the brain as well as to the gastrointestinal system. Intellectual activities can become the predominant ones in the face of physical or social problems (given an environment in which they are valued by someone), and when they become so very early they may often be a defense against ineffectiveness or lack of satisfaction in other spheres. I am not convinced that they need always be defensive but it is evident that a boy who cannot, for some reason (e.g. physical disability, or an immediately older brother) compete effectively in sports can gain at least some status by surpassing the other boys in school work. The need for status is a general one, and certainly not specific to scientists.

There is also a general need for independence, for autonomy, for personal mastery of the environment. The intensity of this varies, probably as a part of the biological makeup of the individual, but it is present in all of us. (I am myself convinced that there is a sex difference, that such needs are stronger in men than in women, but I do not know whether this is a biological or a cultural phenomenon. I am, however, sure that it is related to the paucity of women among scientists.) The intensity also varies because of different experiences in growing up, and perhaps most often in response to greater insecurities the need to develop personal security in this way becomes stronger. That independence is of major importance to these men is very clear. The strength of this need has been a key factor in the motivation that has made them scientists and that has carried them to the top in their professions.

What is its origin? Sometimes one can at least guess. Early loss of a parent can promote deep insecurities. The child who does not fit the group for some reason (even greater intelligence) is insecure beyond most. The overprotected child is insecure when the protection is removed, and even before it is, because he has not been permitted to develop the only security which is satisfying, security within himself.

For reasons which are often obscure, the men who became physical and biological scientists early found special interests and special satisfactions away from personal relations. It was easier for them to become immersed in objects, in things, outside of the human realm. They seem to have less often had intensely close relationships with parents or sibs. I am sometimes not very clear about their relations to their mothers, but they seem much more often than the social scientists to have had fathers whom they respected, but to whom they were not close, and who put relatively few emotional pressures upon them. They were able apparently to achieve a state of masculine identification with their fathers more readily, and this has meant an easier assumption of the masculine role.

Personal pressures on the social scientists were generally

much greater when they were growing up, sometimes from one parent, sometimes from both. That there was not infrequently some confusion in the home over the respective roles of the parents, and very frequent conflict with the fathers seems to have led to difficulties in achieving full masculine status by cultural stereotypes we should long since have outgrown. It seems clear that these emotional problems are related very directly to their choice of profession.

Given insecurities, given the other special situations suggested above, and of course, given a level and appropriate pattern of intellectual abilities, when the boy found that there was a way in which he, personally, could find things out for himself, and that these could be things which mattered to him, it is small wonder that he felt he had found himself. Once these men learned they could do research, once the pleasures of actually accomplishing something on their own had been felt, the choice was made.

That this particular group worked so hard, and has continued to work hard, is strong testimony to the degree of satisfaction they are getting from their activities. Is it then, the case, that to achieve greatly one must be neurotically driven? So far as I can see, while many of these men have greater than the usual insecurities, this is not true of all of them. I will go further and say that I do not see that it need be true. I am sure it is possible for a normally healthy man to become a great scientist,—after all some of these men cannot be otherwise described. But I think it may be harder, in some ways, most particularly in maintaining the degree of concentration needed.

Make no mistake, an extreme degree of concentration is needed. There is so much to learn, so much to master of what is already known, before further steps can be taken, and there are few shortcuts. In this sense, this drivenness is a help to professional accomplishment. But it can become not only a hindrance but a disaster, and not only to a man's personal life, but to his judgment and to his accomplishment in his professional life. Turning to your profession to satisfy other problems, or to refuse to face them, can finally rob you of the pleasure in achievement. There is more than one scientist who reached a high level of achievement, only to be plagued by a bitter depression in the face of high honors. Yet he will maintain, and quite honestly, that he is happy in his work. The trouble is that great as are the satisfactions to be found in work, no profession can supply all you need. Surely something could be done to forestall this development. If our colleges and technical schools could allot a little time to helping a man to know himself and to learn about other aspects of living it would be time well spent.

There are other implications for educational practice in these stories. The discovery that it is possible to find things out for oneself is not a natural part of growing up for every child in our culture. It can be seen clearly in these life histories that for many of these men it was just chance,—the chance, usually, of getting in a class in school where this type of activity was encouraged. Whether it was encouraged because the teacher was genuinely interested in encouraging the children to think for themselves, or whether it was encouraged because the teacher did not want to be bothered with the students and so left them pretty much on their own does not seem to matter too much. The important thing is that they learned that they could satisfy their curiosity by their own efforts. Once they did learn this, good teaching encouraged them, but bad teaching did not stultify them.

It is no easy matter to so design teaching in general that individual thinking is encouraged. For one thing in our public schools our classes are large, and, especially in the grades, there are so many things that must be learned by rote (multiplication tables, reading, spelling) that the atmosphere becomes overwhelmingly one of accepting what is in the book and giving it back unchanged. This carries over strongly to subjects of other sorts, where rote learning actually is not essential, and where it would often be a good idea if the children have an attitude of "it ain't necessarily so" just because the teacher or the book said so. It is chiefly, I think, the carry-over of this authoritarian attitude that is most stultifying. After all, the multiplication table, the forms of spelling, are all conventions, designed for greater convenience in manipulating and conveying ideas, and they should be taught as such, not as basic truths, but I wonder if they ever are. The teachers, themselves, have been brought up by the book and naturally most of them teach as they have been taught. And I must say that the psychologists' invention of the time-saving and effort-saving and hence very popular true and false tests, and similar devices, has not helped matters any in this respect but has probably actually exacerbated the situation.

I mentioned the public schools particularly, but I do not mean to imply that private schools are better in this respect. In fact, with the exception of the anthropologists, very few of this group went to private schools, and I am strongly of the impression that private schools do not produce scientists in any numbers,—they go in much more strongly for the "humanities." Certainly church schools do not produce scientists, even at the college level. Some of the progressive schools have done a good deal, however, to encourage children to follow out their own pursuits. Unfortunately a good many of them have carried this so far that they have neglected to insist upon the children also learning the conventions which are really necessary.

Let me make it clear that I do not think the major function of any school is to produce scientists. The major function of our schools is to aid in the production of good citizens. It is true, I think, that scientists are usually very good citizens,— they mind their own business, they pay at least as much attention to civic duties as the average man does, they do not enrich themselves at others' expense, they and their families rarely become public charges, and the more violent crimes are practically unknown among them. You will remember that even on test material they show up as an unaggressive group, on the whole. (The scientists involved in espionage have been very few, indeed, and misguided as they may have been, they have acted on principle and not for personal gain.) But obviously scientists are not the only good citizens. The point that is important here, and much more important than the production of scientists, is that the things that make good scientists are also the things that make good citizens, in a democracy. This is true only for democratic forms of government. A democracy can only exist effectively if the citizens are able to participate freely and intelligently. If they are not free to think and to feel, they are not free. We give a good deal of lip service to the concept of an enlightened electorate, but we are often still not quite sure that people can be trusted. There is a good deal of evidence in the clinical literature that people, freed of neurotic needs and in a position to make a really conscious choice, will not choose asocially. We are more hampered by our own fear of ourselves, and of our nature than we are by anything else.

I do not want to imply, either, that the job of encouraging people to do their own thinking is entirely the job of the schools. Far from it. A particular charge does rest upon them because they are our most nearly universal institution, the one with which most people in our culture have at least some contact. This is not nearly so true of any other institution (except the family insofar as that can be called an institution). Our churches seem not to accept a similar charge. It is a very rare church indeed that encourages its members to think for themselves in religious matters, or even tolerates this, and in most of them the clergy are quite ready to lay down the law in other fields also. Now it is possible to be a devout member of any church and still do some individual thinking in other matters, but it is something of a trick, and does not make for the healthiest sort of adjustment. It is also possible to accept the tenets of any church on the basis of one's own individual thinking I suppose, if only because they were developed by people in the first place. The difficulty there is the cultural changes which we have undergone since any of these doctrines were first enunciated, and the tenacious resistance to any change, even in the form of a doctrine, although the spirit may remain unchanged. And, of course, so long as you believe that man is essentially evil in nature, and a more vicious doctrine was never promulgated, it follows that he is often going to need to have his ears slapped back, and who should do this but the clergy? In all my years of clinical experience, even on the back wards of mental hospitals, I have seen no evidence for the doctrine of original sin, and its consequent requirement of salvation. Quite the contrary. Man has evolved into what he is and his past is as old as the world. His future is in his own hands, and this has not been true in the same way of any other creature. He has evolved into a responsible being, into one with the power of choice, and this he cannot evade and cannot delegate. It is when he tries to do so, and thus denies his nature, that breakdowns occur. His only hope is to maintain freedom of choice. If he can do this, if he will accept fully what he is, the future will take care of itself.

Freedom, like charity, begins at home. So does the disciplining necessary for social living. It is the achievement of a balance between restraint and freedom, and the final internalizing of the minimum restraints that are necessary, the taking over of them by the individual himself, that is the hallmark of maturity. More than anything else the family situation can help or hinder this. Even so, I do not believe that the hindrance need ever be insuperable or that any situation is irremediable. Look at the evidence here. Some of these subjects had difficult childhoods because of illness or poverty or parental death, but they have gone on to great achievements in spite of this, or even perhaps because of this.

Indeed, it would seem that a completely placid life, although not necessarily a hindrance to development, is also not necessarily an advantage. It is true that just the exigencies of growing up are likely to introduce a certain minimum number of difficulties, and that we hardly need to go out of our way to introduce others. Nevertheless there is a strong tendency in our culture to reduce difficulties, to take the attitude that we do not want our children to go through what we had to go through, to try to protect them from many normal vicissitudes. I do not deprecate this altogether, and have certainly been guilty of it myself, but at the same time one unfortunate result of it seems to be the postponing of full personal responsibility, economic as well as otherwise. Many of these men achieved this relatively early, in spite of their long schooling. There are very great difficulties in the way of developing full responsibility in other matters without full economic responsibility, and these are becoming increasingly crucial as more and more of our population continue their schooling to higher ages.

The effect of being in a situation which permits or requires individual action has been strikingly demonstrated in diverse situations, as for example, in factories where it has been shown that groups given a chance to organize the routine work themselves are more productive than groups where the system is laid down by authority. This, I think, is just part of the nature of man. These particular men were fortunate in finding a field in which they could achieve the greatest amount of this and in being able to make a profession of it.

But you do not have to be a scientist to experience this sort of satisfaction. Nor do you have to make a profession of science to develop scientific attitudes, which will make you a better and a happier citizen. Research in the broadest sense is more a habit of mind and a method of approach to problems than a specific technique. Certainly there is nothing esoteric about it (as I hope this book has demonstrated about clinical psychological research, at least). You can develop this sort of attitude about anything you do, and have more fun doing it. (You may run into difficulties if you try it on your job too suddenly,—remember the story of the physicist trying to be a salesman.) It is not always easy. You must be free, first, free to observe and free to follow where your ob-

243

servations lead you, even if it means discarding some cherished beliefs. You must be patient. You must learn to wait until enough evidence is in. You must be willing to start at the beginning and do things all over again. Above all, you must be willing to see that you can be wrong, even if that means that your most cherished rival is right.

The gardener who adds some preparation to part of his soil and watches to see how the results compare with a plot that has not had the preparation is doing research. The more systematically he does this and the more careful his records, the better the results he is likely to get. The housewife who experiments with a recipe until she gets the finished product just right is doing research.

Is it worth the bother? That is up to you. But it has advantages you may not have thought of. I have said that you must have a measure of freedom to take a scientific attitude at all. It is also true that the more you take it the more freedom you will have. Freedom breeds freedom. Nothing else does.